A Comprehensive Guide to Information Security Management and Audit

This text is written to provide readers with a comprehensive study of information security and management system, audit planning and preparation, audit techniques and collecting evidence, international information security (ISO) standard 27001, and asset management. It further discusses important topics such as security mechanisms, security standards, audit principles, audit competence and evaluation methods, and the principles of asset management. It will serve as an ideal reference text for senior undergraduate and graduate students and researchers in fields including electrical engineering, electronics and communications engineering, computer engineering, and information technology.

This book explores information security concepts and applications from an organizational information perspective and explains the process of audit planning and preparation. It further demonstrates audit techniques and collecting evidence to write important documentation by following the ISO 27001 standards.

This book:

- Elaborates on the application of confidentiality, integrity, and availability (CIA) in the area of audit planning and preparation.
- Covers topics such as managing business assets, agreements on how to deal with business assets, and media handling.
- Demonstrates audit techniques and collects evidence to write important documentation by following the ISO 27001 standards.
- Explains how the organization's assets are managed by asset management and access control policies.
- Presents seven case studies.

A Comprehensive Guide to Information Security Management and Audit

Rajkumar Banoth, Narsimha Gugulothu and
Aruna Kranthi Godishala

CRC Press is an imprint of the
Taylor & Francis Group, an **informa** business

First edition published 2023
by CRC Press
6000 Broken Sound Parkway NW, Suite 300, Boca Raton, FL 33487-2742

and by CRC Press
4 Park Square, Milton Park, Abingdon, Oxon, OX14 4RN

CRC Press is an imprint of Taylor & Francis Group, LLC

© 2023 Rajkumar Banoth, Narsimha Gugulothu and Aruna Kranthi Godishala

Reasonable efforts have been made to publish reliable data and information, but the author and publisher cannot assume responsibility for the validity of all materials or the consequences of their use. The authors and publishers have attempted to trace the copyright holders of all material reproduced in this publication and apologize to copyright holders if permission to publish in this form has not been obtained. If any copyright material has not been acknowledged please write and let us know so we may rectify in any future reprint.

Except as permitted under U.S. Copyright Law, no part of this book may be reprinted, reproduced, transmitted, or utilized in any form by any electronic, mechanical, or other means, now known or hereafter invented, including photocopying, microfilming, and recording, or in any information storage or retrieval system, without written permission from the publishers.

For permission to photocopy or use material electronically from this work, access www.copyright.com or contact the Copyright Clearance Center, Inc. (CCC), 222 Rosewood Drive, Danvers, MA 01923, 978-750-8400. For works that are not available on CCC please contact mpkbookspermissions@tandf.co.uk

Trademark notice: Product or corporate names may be trademarks or registered trademarks and are used only for identification and explanation without intent to infringe.

Library of Congress Cataloging-in-Publication Data
Names: Banoth, Rajkumar, author. | Gugulothu, Narsimha, author. | Godishala, Aruna Kranthi, author.
Title: A comprehensive guide to information security management and audit / [edited by] Rajkumar Banoth, Narsimha Gugulothu and Aruna Kranthi Godishala.
Description: First edition. | Boca Raton : CRC Press, 2023. | Includes bibliographical references and index.
Identifiers: LCCN 2022016803 (print) | LCCN 2022016804 (ebook) | ISBN 9781032344430 (hardback) | ISBN 9781032344478 (paperback) | ISBN 9781003322191 (ebook)
Subjects: LCSH: Confidential business information--Security measures. | Computer networks--Security measures. | Information audits. | Computer security.
Classification: LCC HD38.7 .B36 2023 (print) | LCC HD38.7 (ebook) | DDC 658.4/7--dc23/eng/20220711
LC record available at https://lccn.loc.gov/2022016803
LC ebook record available at https://lccn.loc.gov/2022016804

ISBN: 9781032344430 (hbk)
ISBN: 9781032344478 (pbk)
ISBN: 9781003322191 (ebk)

DOI: 10.1201/9781003322191

Typeset in Sabon
by Deanta Global Publishing Services, Chennai, India

The dedication of this book is to

My loveliest daughter
Aadhya Kruthi

And to you
If you have been struck with this book until the very end.

Contents

Author Bios   xv
Preface   xvii
Acknowledgments   xix
Acronyms/Abbreviations   xxi

1 Information Security and Management System   **1**

    Information Security Overview 1
    1.1   *The OSI Security Architecture 1*
    1.2   *Information Security 2*
        Security attacks 2
            Passive attack 2
            Active attack 3
    1.3   *Security Services 4*
        Confidentiality 4
        Authentication 5
        Integrity 5
        Non-repudiation 6
        Access control 6
        Availability 6
    1.4   *Security Mechanisms 6*
        Specific security mechanisms 6
        Pervasive security mechanisms 7
        Model for network security 8
        Some basic terminologies 9
        Cryptography 9
        Cryptanalysis 9
        Introduction and importance of Information
            Security and Management System (ISMS) 10
            Why security management? 10

1.5   The CIA and DAD Triads 11
      The CIA triad 11
      The DAD triad 12
            How are the CIA and DAD triads
                  mutually exclusive? 12
            How can you relate the CIA triad
                  in your everyday life? 12
1.6   ISMS Purpose and Objectives 13
      Introduction to information security policies 13
      Elements of information security policy 13
      Scope (objective) 14
      Security policies 14
      Security policy development 16
      Phased approach 16
      Security policy contributors 16
      Security policy audience 18
      Policy categories 19
1.7   Frameworks 19
      Policy categories 20
      Additional regulations and frameworks 20
      Security management policies 21
1.8   Security Standards 23
      Security standard example 23
1.9   Standard 24
      Services 24
      Initial password and login settings 24
      Send mail 25
1.10 Security procedures 25
      Security procedure example 25
      Apache web server security procedure 25
1.11 Security Guidelines 27
      Security guideline example 27
      Password selection guidelines 27
            Do 27
            Don't 28
            Suggestions 28
1.12 Compliance vs. Conformance 28
      Compliance 28
      Conformance 29
      Special applications 29
      Conclusion on compliance and conformance 30
      Bibliography 30

2 Audit Planning and Preparation 33

Introduction 33
2.1　Reasons for Auditing 33
2.2　Audit Principles 34
　　2.2.1　Planning 34
　　2.2.2　Honesty 35
　　2.2.3　Secrecy 35
　　2.2.4　Audit evidence 35
　　2.2.5　Internal control system 35
　　2.2.6　Skill and competence 35
　　2.2.7　Work done by others 35
　　2.2.8　Working papers 35
　　2.2.9　Legal framework 36
　　2.2.10　Audit report 36
2.3　Process of Audit Program Management 36
　　2.3.1　Preparing for an audit 37
　　2.3.2　Audit process 37
2.4　Audit competence and evaluation methods 38
　　2.4.1　Audit of individuals 39
　　2.4.2　Audit of sole trader's books of accounts 39
　　2.4.3　Audit of partnership firm 39
　　　　　Important provision of Partnership Act 39
　　2.4.4　Government audit 40
　　　　　Important features of the government
　　　　　　audit 40
　　　　　Objectives 40
　　2.4.5　Statutory audit 41
　　2.4.6　Audit of companies 41
　　2.4.7　Audit of trust 41
　　2.4.8　Audit of cooperative societies 42
　　2.4.9　Audit of other institutions 42
　　　　　Cost audit 42
　　　　　Objectives of cost audit 42
　　2.4.10　Tax audit 42
　　2.4.11　Balance sheet audit 43
　　　　　Continuous audit 43
　　　　　Annual audit 43
　　2.4.12　Partial audit 44
　　2.4.13　Internal audit 44
　　2.4.14　Management audit 44
　　　　　Objectives of management audit 44
　　2.4.15　Post & Vouch Audit 45

       2.4.16 Audit in depth 45
       2.4.17 Interim audit 45
  2.5 Audit Responsibilities 46
       2.5.1 Reporting on the financial statements 48
       2.5.2 Unmodified opinions 48
       2.5.3 Modified opinions 49
       2.5.4 Emphasizing certain matters
              without modifying the opinion 49
       2.5.5 Communicating "other matters" 50
       2.5.6 Other information included in the
              annual report 50
       2.5.7 Other legal and regulatory requirements 50
       2.5.8 Reporting on the financial statements 50
  2.6 Audit Time and Process Flow 52
       2.6.1 What is a process? 52
       2.6.2 Process description 52
       2.6.3 Control of processes 55
       2.6.4 Advanced process and system modeling 57
  2.7 ISMS audit checklist 57
       2.7.1 Why ISO 27001 Checklist is required? What
              is the importance of ISO 27001 Checklists? 57
       2.7.2 Who all can use ISO 27001 Audit Checklist? 58
       2.7.3 How many ISO 27001 Checklists are available? 59
       2.7.4 How to find out which ISO 27001
              Checklists are suitable for me? 59
              2.7.4.1 For an organization aiming
                       for ISO 27001 Certification 59
              2.7.4.2 For a head of the department? 60
              2.7.4.3 For a CISO (Chief Information
                       Security Officer) 60
              2.7.4.4 For a CTO (Chief Technology
                       Officer) and CIO 60
              2.7.4.5 For IT department professionals 60
              2.7.4.6 For preparing for a job interview 60
       2.7.5 Important information on ISO 27001
              Checklist file 60
       2.7.6 Who has prepared and who has
              validated ISO 27001 Checklists? 61
       2.7.7 What is the basis of the ISO 27001
              Checklist? 61
       2.7.8 How to use ISO 27001 Checklist? 61
       Bibliography 61

3 Audit Techniques and Collecting Evidence 63

- 3.1 Auditor Quality and Selection 63
  - How to prepare for an auditor selection process 63
  - Four steps to select an auditor 64
- 3.2 Audit Script 66
  - Customizing audit scripts 67
  - Customize standard audit scripts 67
  - To customize an audit script 67
  - Using standard audit scripts 68
  - Create new audit scripts 70
  - Enable audit scripts 74
  - Install audit scripts 76
  - Print audit scripts 79
  - Remove audit script 80
  - Set audit scripts 80
  - Update audit scripts 83
  - Using product-specific audit scripts 84
- 3.3 Audit Stages 85
  - Levels of audit engagement 86
- 3.4 Audit Techniques 86
  - Inspection 86
  - Observation 87
  - Inquiry and confirmation 87
  - Computation 88
  - Analytical procedures 88
- 3.5 Collecting Evidence through Questions 88
  - Inquiry 89
  - Sufficient appropriate audit evidence 89
  - Ways of collecting audit evidence 89
    - Inspection 89
    - Observation 90
    - External confirmation 90
    - Documentation 90
    - Recalculation 90
    - Re-performance 90
    - Analytical procedures 91
    - Inquiry 91
- 3.6 Observation 91
- 3.7 Reporting to Audit Finding 91
  - Different types of audit findings 92
  - Respond to audit findings 93

3.8 Audit Team Meeting 94
 Importance of opening meetings 94
 Opening meeting 94
 Introduction 94
 Confirm the scope and objectives of the assessment 94
 Confirm communications, resources, and escorts 95
 Current number of employees 95
 Confirm auditor confidentiality 95
 Explain the audit program and the reporting
   process for deficiencies 95
 Confirm time and place for closing meeting 95
 Appeals process 96
 Audit team safety induction 96
3.9 Nonconformities and Observation 96
 Example of a well-written nonconformity 98
 Auditors are held to a higher standard 98
3.10 Corrective and Preventive Actions 99
 An in-depth look at corrective and preventive action 100
 Corrective action 100
 What's the scope of corrective action? 101
 Benefits of corrective action 101
 Issues of corrective action 101
 Corrective Action Request (CAR) 101
 Preventive action 102
 What's the scope of preventive action? 102
 How does corrective action differ from preventive action? 102
 How is corrective action similar to preventive action? 103
 Corrective action and preventive action in practice 103
 Implementing corrective and preventive action 104
 Using the corrective and preventive action subsystem 104
 Bibliography 105

4 ISO 27001   107

4.1 Overview of an Information Security
 and Management System 107
 ISO publishes two standards that focus
   on an organization's ISMS: 108
4.2 Purchase a Copy of the ISO/IEC Standards 110
4.3 Determine the Scope of the ISMS 112
4.4 Identify Applicable Legislation 113
 Scope and purpose 113

4.5 Define a Method of Risk Assessment 114
4.6 Create an Inventory of Information Assets to Protect 115
4.7 Identify Risks 116
4.8 Assess the Risks 117
4.9 Identify Applicable Objectives and Controls 118
4.10 Set Up Policy, Procedures, and Documented Information to Control Risks 122
4.11 Allocate Resources and Train the Staff 123
4.12 Monitor the Implementation of the ISMS 124
4.13 Prepare for the Certification Audit 125
     Bibliography 126

5 Asset Management 127

5.1. What Are Assets According to ISO 27001? 127
5.2. Why Are Assets Important for Information Security Management? 127
5.3. How to Build an Asset Inventory? 128
5.4. Who Should be the Asset Owner? 128
5.5. ISO 27001/ISO 27005 Risk Assessment & Treatment – Six Basic Steps 129
5.6. The Basic Steps Will Shed Light on What One Has to Do 129
      5.6.1 ISO 27001 risk assessment methodology 129
      5.6.2 Risk assessment implementation 129
      5.6.3 Risk treatment implementation 129
      5.6.4 ISMS risk assessment report 130
      5.6.5 Statement of applicability 130
      5.6.6 Risk treatment plan 130
5.7. ISO 27001 Controls from Annex A 130
      5.7.1 How many domains are there in ISO 27001? 131
      5.7.2 What are the 14 domains of ISO 27001? 131
5.8. The Importance of Statement of Applicability for ISO 27001 132
      5.8.1 Why it is needed? 133
5.9. ISO 27001: A.8 Asset Management 134
      5.9.1 Introduction 134
      5.9.2 Level of assets 135
      5.9.3 Asset management 136
      5.9.4 The principles of asset management 139
      5.9.5 Asset life cycle 140
             How to go about it? 142
      5.9.6 Seven steps to implement asset management 143

5.10. Responsibility for Assets 144
    A.8.1   Responsibility for assets 144
        A.8.1.1   Inventory of assets 144
        A.8.1.2   Ownership of assets 145
        A.8.1.3   Acceptable use of assets 145
        A.8.1.4   Return of asset 146
        A.8.1.5   Responsibility for assets 146
5.11. Information Classification 153
    A.8.2   Information classification 153
        A.8.2.1   Classification of information 153
        A.8.2.2   Labeling of information control 154
        A.8.2.3   Handling of assets 155
5.12. Media Handling 158
    A.8.3   Media handling 158
        A.8.3.1   Management of removable media 158
        A.8.3.2   Disposal of media 159
        A.8.3.3   Physical media transfer 160
5.13. BYOD 161
    5.13.1   What are the types of BYOD? 161
    5.13.2   Why is BYOD important? 161
    5.13.3   Benefits of BYOD improve productivity 162
        Boost employee satisfaction 162
        Cut enterprise costs 162
        Attract new hires 162
    5.13.4   Risks of BYOD 162
    5.13.5   Keys to effective BYOD 163
    5.13.6   Guidelines to help plan and implement effective BYOD 163
    Bibliography 164

Index 167

Author Bios

Rajkumar Banoth, BTech, MTech, PhD, is Associate Professor at Marwadi University, Rajkot, India, IEEE senior member, and Cyber Security Operations Certified Trainer. He has published six textbooks comprising the domains Networking, Computer Organization and Architecture, and Computer Forensic; published in 18 Hi-indexed SCI and Scopus journals; and presented 9 conference papers. He also has a membership with "The Institution of Engineers (India)".

Narsimha Gugulothu, BTech, MTech, PhD, is currently working as Vice Principal and Professor in the Department of Computer Science and Engineering at JNTUH College of Engineering Sultanpur, Telangana, India. He is an applauded academician and researcher possessing 22 years of experience in teaching, research, and administration in well-reputed educational institutions majorly in state universities JNTU-Kakinada and JNTU-Hyderabad alongside private institutions.

Aruna Kranthi Godishala, BTech, MTech, is a Research Scholar at Brunei University Darussalam, Brunei, and obtained a BTech in Computer Science and Engineering and an MTech in Software Engineering from Jawaharlal Nehru Technological University, Hyderabad, India.

Preface

It is important to note that securing and keeping information from parties who do not have the authorization to access such information is an extremely important issue. To address this issue, it is essential for an organization to implement an ISMS (Information Security and Management System) standard such as ISO 27001 to address the issue comprehensively. The authors of this volume have explained various aspects of the security framework.

In addition, ISMS does not only help organizations to assess their information security compliance with ISO 27001 but can also be used as a monitoring tool, helping organizations monitor the security statuses of their information resources and potential threats. ISMS is developed to provide solutions to solve obstacles, difficulties, and expected challenges associated with literacy and governance of ISO 27001. It also covers functions to assess the risk level of organizations toward compliance with ISO 27001.

The main objective of this volume is (1) to overcome security threats which are exposed through organizations and their information systems and networks and (2) to reduce risk effects which undermine or create difficulties for its business operations.

Therefore, there is a need for an information security management methodology to protect information systematically. This creates the importance of ISMS.

The information provided here will act as blueprints for managing information security within business organizations. After completion of this volume, one will be able to understand key elements of ISO 27001, etc. standards, understand key information security issues, plan an audit against a set of audit criteria, successfully execute an information security management system audit, create clear, concise, and relevant audit reports, and communicate the audit findings to a client.

Acknowledgments

Writing this book, entitled A Comprehensive Guide to Information Security Management and Audit, is not soft to touch as I thought and more important than I could have ever imagined.

I'm eternally grateful to my Dean Dr. Rajendrasinh Bahadursinh Jadeja, Marwadi University, who took extra interest to motivate me to write this book where he didn't have to. He taught me discipline, manners, respect, and so much more that has helped me succeed in my current tenure at Marwadi University.

To my mother (Ullamma), and my brother's son Bhanuprakash alias Chitti:

Thank you for letting me know that you have nothing but great memories of me and special thanks to chitti for the timely design of images for this book.

To aunt and uncle Subhashini, a teacher by profession, and Dr. Hanumantha Rao, a doctor by profession for always being the persons I could turn to during those dark and tough years. They sustained me in ways that I never knew I needed.

Finally, to all those who have been a part of my journey to write this book: Miss. Kajalben Kalubhai Tanchak, Mr. Alla Poorna Chandra Reddy.

To the CRC Press | Taylor & Francis Group team Marc Gutierrez, Gauravjeet Singh Reen, and Isha Ahuja.

Acronyms/Abbreviations

Acronyms/Abbreviations	Description
CIA	Confidentiality Integrity Availability
DAD	Disclosure Alteration Destruction
ISMS	Information Security Management System
ISO	International Organization for Standardization
IEC	International Electrotechnical Commission
CISO	Chief Information Security Officer
CTO	Chief Technology Officer
CIO	Chief Information Officer
BYOD	Bring-Your-Own-Device

Chapter 1

Information Security and Management System

INFORMATION SECURITY OVERVIEW

Computer data often travels from one computer to another, leaving the safety of its protected physical surroundings. Once the data is out of hand, people with bad intentions could modify or forge your data, either for amusement or for their own benefit. Cryptography can reformat and transform our data, making it safer on its trip between computers. The security is based on the essentials of secret codes, augmented by modern mathematics that protects our data in powerful ways. Few security types are defined as follows:

Information security: Practice of protecting information by mitigating information risks.
Computer security: Generic name for the collection of tools designed to protect data and thwart hackers.
Network security: Measures to protect data during its transmission.
Internet security: Measures to protect data during its transmission over a collection of interconnected networks.

1.1 THE OSI SECURITY ARCHITECTURE

To assess effectively the security needs of an organization and to evaluate and choose various security products and policies, the manager responsible for security needs in a systematic way of defining the requirements for security and characterizing the approaches to satisfying those requirements. The OSI security architecture provides a useful, abstract overview of many of the concepts. The OSI security architecture mainly focuses on security attacks, mechanisms, and services.

1.2 INFORMATION SECURITY

It can be defined as "measures adopted to prevent the unauthorized use, misuse, modification, or denial of use of knowledge, facts, data or capabilities". Three aspects of IS are as follows:

Security attack: Any action that comprises the security of information.
Security service: It is a processing or communication service that enhances the security of the data processing systems and information transfer. The services are intended to counter security attacks by making use of one or more security mechanisms to provide the service.
Security mechanism: A mechanism that is designed to detect, prevent, or recover from security.

Security attacks

These attacks involve any action that compromises the security of information owned by an organization. Information security is about how to prevent attacks or, failing that, to detect attacks on information-based systems. Often threat and attack are used to mean the same thing. But that is not always true; for details, see Table 1.1. There are a wide range of attacks. In this, we will mainly focus on generic types of attacks. These are two, as follows:

- Passive attack.
- Active attack.

For more categorization of attack, refer Figure 1.1.

Passive attack

A passive attack attempts to learn or make use of information from the system but does not affect system resources. The illustration of passive attack is given in Figure 1.2. Two types of passive attack are as follows:

- Release of message content: It may be desirable to prevent the opponent from learning the contents (i.e., sensitive or confidential info) of the transmission.
- Traffic analysis: It is a more subtle technique where the opponent could determine the location and identity of communicating hosts and could observe the frequency and length of encrypted messages being exchanged, thereby guessing the nature of communication taking place. Passive attacks are very difficult to detect because they do not involve any alteration of the data. As the communications take place in a very normal fashion, neither the sender nor receiver is aware that a third party has read the messages or observed the traffic pattern. So

Information Security and Management System 3

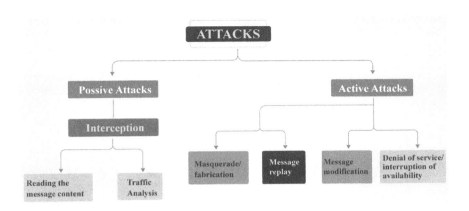

Figure 1.1 Categorization of attack.

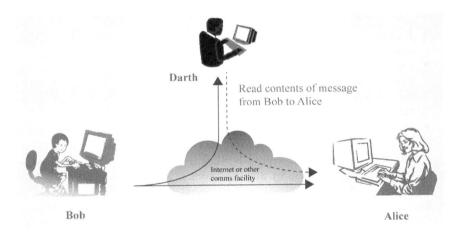

Figure 1.2 Passive attack.

the emphasis in dealing with passive attacks is on prevention rather than detection.

Active attack

Active attacks involve some modification of the data stream or the creation of a false stream. For a pictorial understanding of the active attack, refer to Figure 1.3. An active attack attempts to alter system resources or affect their operation. Four types of active attack are as follows:

- Masquerade: Here, an entity pretends to be some other entity. It usually includes one of the other forms of active attack.

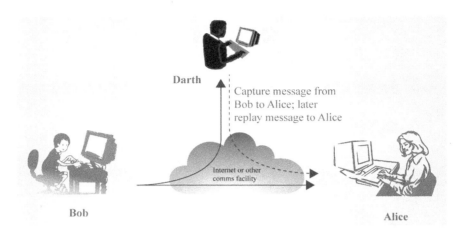

Figure 1.3 Active attack.

- Replay: It involves the passive capture of a data unit and its subsequent retransmission to produce an unauthorized effect.
- Modification of messages: It means that some portion of a legitimate message is altered or that messages are delayed to produce an unauthorized effect. Ex: "John's acc no is 2346" is modified as "John's acc no is 7892".
- Denial of service: This attack prevents or inhibits the normal use or management of communication facilities. Ex: a: Disruption of the entire network by disabling it, b: Suppression of all messages to a particular destination by a third party.

Active attacks present the opposite characteristics of passive attacks. Whereas passive attacks are difficult to detect, measures are available to prevent their success. On the other hand, it is quite difficult to prevent active attacks absolutely because of the wide variety of potential physical, software, and network vulnerabilities. Instead, the goal is to detect active attacks and recover.

1.3 SECURITY SERVICES

It is a processing or communication service that is provided by a system to give a specific kind of protection to system resources. Security services implement security policies and are implemented by security mechanisms.

Confidentiality

Confidentiality is the protection of transmitted data from passive attacks. It is used to prevent the disclosure of information to unauthorized individuals or systems. It has been defined as "ensuring that information is accessible only to those authorized to have access". The other aspect of confidentiality is the protection of traffic flow from analysis. Ex: A credit card number has to be secured during an online transaction.

Authentication

This service assures that a communication is authentic. For a single message transmission, its function is to assure the recipient that the message is from the intended source. For an ongoing interaction, two aspects are involved. First, during connection initiation, the service assures the authenticity of both parties. Second, the connection between the two hosts is not interfered allowing a third party to masquerade as one of the two parties. Two specific authentication services defined in X.800 are as follows:

- Peer entity authentication: It verifies the identities of the peer entities involved in communication. It is useful at the time of connection establishment and during data transmission. It provides confidence against a masquerade or a replay attack.
- Data origin authentication: It assumes the authenticity of the source of the data unit but does not provide protection against duplication or modification of data units. It supports applications like electronic mail, where no prior interactions take place between communicating entities.

Integrity

Integrity means that data cannot be modified without authorization. Like confidentiality, it can be applied to a stream of messages, a single message, or selected fields within a message. Two types of integrity services are available. They are as follows:

- Connection-oriented integrity service: This service deals with a stream of messages and assures that messages are received as sent, with no duplication, insertion, modification, reordering, or replays. Destruction of data is also covered here. Hence, it attends to both message stream modification and denial of service.
- Connectionless-oriented integrity service: It deals with individual messages regardless of a larger context, providing protection against message modification only.

An integrity service can be applied with or without recovery. Because it is related to active attacks, the major concern will be detection rather than prevention. If a violation is detected and the service reports it, either human intervention or automated recovery machines are required to recover.

Non-repudiation

Non-repudiation prevents either sender or receiver from denying a transmitted message. This capability is crucial to e-commerce. Without it, an individual or entity can deny that he, she, or it is responsible for a transaction and, therefore, not financially liable.

Access control

This refers to the ability to control the level of access that individuals or entities have to a network or system and how much information they can receive. It is the ability to limit and control access to host systems and applications via communication links. For this, each entity trying to gain access must first be identified or authenticated so that access rights can be tailored to the individuals.

Availability

It is defined as the property of a system or a system resource being accessible and usable upon demand by an authorized system entity. The availability can significantly be affected by a variety of attacks, some are amenable to automated counter measures, i.e., authentication and encryption, and others need some sort of physical action to prevent or recover from the loss of availability of elements of a distributed system.

1.4 SECURITY MECHANISMS

According to X.800, the security mechanisms are divided into those implemented in a specific protocol layer and those that are not specific to any particular protocol layer or security service. X.800 also differentiates reversible and irreversible encipherment mechanisms. A reversible encipherment mechanism is simply an encryption algorithm that allows data to be encrypted and subsequently decrypted, whereas irreversible encipherment includes hash algorithms and message authentication codes used in digital signature and message authentication applications.

Specific security mechanisms

These are incorporated into the appropriate protocol layer in order to provide some of the OSI security services.

> Encipherment: It refers to the process of applying mathematical algorithms for converting data into a form that is not intelligible. This depends on the algorithm used and encryption keys.

Digital signature: It is the appended data or a cryptographic transformation applied to any data unit allowing to prove the source and integrity of the data unit and protect against forgery.

Access control: It involves a variety of techniques used for enforcing access permissions to the system resources.

Data integrity: It involves a variety of mechanisms used to assure the integrity of a data unit or stream of data units.

Authentication exchange: It is a mechanism intended to ensure the identity of an entity by means of information exchange.

Traffic padding: The insertion of bits into gaps in a data stream to frustrate traffic analysis attempts is called traffic padding.

Routing control: It enables the selection of particular physically secure routes for certain data and allows routing changes once a breach of security is suspected.

Notarization: The use of a trusted third party to assure certain properties of data exchange is called notarization.

Pervasive security mechanisms

These are not specific to any particular OSI security service or protocol layer. Refer to Figure 1.4 for further understanding.

Trusted functionality: It is perceived to be correct with respect to some criteria.

Security level: It is the marking bound to a resource (which may be a data unit) that names or designates the security attributes of that resource.

Event detection: It is the process of detecting all the events related to network security.

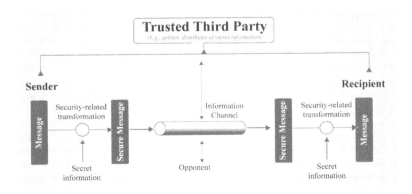

Figure 1.4 Model for network security.

Security audit trail: Data is collected and potentially used to facilitate a security audit, which is an independent review and examination of system records and activities.

Security recovery: It deals with requests from mechanisms, such as event handling and management functions, and takes recovery actions.

Model for network security

Data is transmitted over a network between two communicating parties, who must cooperate for the exchange to take place. A logical information channel is established by defining a route through the internet from source to destination by the use of communication protocols by the two parties. Whenever an opponent presents a threat to confidentiality, the authenticity of the information, security aspects come into play. Two components are present in almost all the security-providing techniques.

A security-related transformation of the information should be sent, making it unreadable by the opponent, and the a code should be added based on the contents of the message, to verify the identity of the sender. Some secret information should be shared by the two principals, and, it is hoped, unknown to the opponent. An example is an encryption key used in conjunction with the transformation to scramble the message before transmission and unscramble it on reception.

A trusted third party may be needed to achieve secure transmission. It is responsible for distributing the secret information to the two parties, while keeping it away from any opponent. It also may be needed to settle disputes between the two parties regarding the authenticity of a message transmission. The general model shows that there are four basic tasks in designing a particular security service:

- Design an algorithm for performing the security-related transformation. The algorithm should be such that an opponent cannot defeat its purpose.
- Generate the secret information to be used with the algorithm.
- Develop methods for the distribution and sharing of secret information.
- Specify a protocol to be used by the two principals that makes use of the security algorithm and the secret information to achieve a particular security service.

Various other threats to information systems like unwanted access still exist. The existence of hackers attempting to penetrate systems accessible over a network remains a concern. Another threat is the placement of some logic in computer systems affecting various applications and utility programs. This inserted code presents two kinds of threats.

Some basic terminologies

- Cipher text: The coded message.
- Cipher: Algorithm for transforming plain text to cipher text.
- Key: Info used in cipher known only to sender/receiver.
- Encipher (Encrypt): Converting plain text to cipher text.
- Decipher (Decrypt): Recovering cipher text from plain text.
- Cryptography: Study of encryption principles/methods.
- Cryptanalysis (Codebreaking): The study of principles/methods of deciphering cipher text without knowing the key.
- Cryptology: The field of both cryptography and cryptanalysis.

Cryptography

Cryptographic systems are generally classified along three independent dimensions:

Type of operations used for transforming plain text to cipher text:

All the encryption algorithms are based on two general principles, substitution, in which each element in the plaintext is mapped into another element, and transposition, in which elements in the plain text are rearranged.

The number of keys used:

If the sender and receiver use the same key, then it is said to be a symmetric key (or) single key (or) conventional encryption. If the sender and receiver use different keys, then they are said to be public key encryption.

The way in which the plain text is processed:

A block cipher processes the input and block of elements at a time, producing an output block for each input block. A stream cipher processes the input elements continuously, producing output elements one at a time, as it goes along.

Cryptanalysis

The process of attempting to discover X, K, or both is known as cryptanalysis. The strategy used by the cryptanalysis depends on the nature of the encryption scheme and the information available to the cryptanalyst. There are various types of cryptanalytic attacks based on the amount of information known to the cryptanalyst.

> Cipher text only: A copy of cipher text alone is known to the cryptanalyst.
> Known plaintext: The cryptanalyst has a copy of the cipher text and the corresponding plaintext.
> Chosen plaintext: The cryptanalysts gain temporary access to the encryption machine. They cannot open it to find the key; however, they can encrypt a large number of suitably chosen plain texts and try to use the resulting cipher texts to deduce the key.

Chosen cipher text: The cryptanalyst obtains temporary access to the decryption machine, uses it to decrypt several strings of symbols, and tries to use the results to deduce the key.

Introduction and importance of Information Security and Management System (ISMS)

Why security management?

We have introduced how cryptography, security protocols, and system security techniques allow us to build more secure systems and networks. But is that all we require to secure an organization? Are technical systems enough to ensure that the organization and data are going to be secure? Most of the time, when we think about security, we think about the technologies we need to use for providing security for information, either at rest or in transit, or other information systems, such as computers and networks. Unfortunately, security is not only about the technical means to achieve it but also about the processes and people involved in those processes.

It doesn't matter if we encrypt some data using the most secure encryption algorithm if we use a very simple password. In the same sense, it doesn't matter if your organization buys the most expensive and advanced security software if the employees can uninstall it from their computers at their will. Security management is concerned with how to use security technologies in the real world to protect organizational assets. This means putting together the technical aspects of security, processes, and people so that organizations can achieve their business goals.

Security management is not only deciding which security technology to use. Security controls need to be configured, integrated into the organization, monitored, updated, and replaced as necessary. Security technology that is not properly used won't help to protect organization assets. Security management covers all aspects that help an organization to preserve the three famous security goals. These are, as you probably already know, confidentiality, integrity, and availability. In the context of security management, confidentiality means that information assets should only be read by those users that are entitled to do so. Integrity is about preventing users from modifying organizational assets when they do not have the necessary authorization.

Finally, availability means that organization assets can be accessed by authorized users when needed. Security management also provides accountability and auditability and serves to put compliance to standards and regulations. In today's world, there are many ways in which the current regulations affect the controls we need to establish to protect information. For example, in European countries, data protection legislation requires all holders of personal identifiable information to protect it appropriately. Information security management involves staff management, too. In fact, a staff is typically the biggest security risk and also the most important security control.

Security management activities related to staff include initial vetting of new employees, security training, and awareness, among others. Security training is meant to encourage the staff to follow good security practices and follow the organization's security policies. Obviously, the success of these initiatives varies hugely and depends on the approach followed to train employees on security practices. Forcing employees to follow static online courses doesn't help employees realize the importance of security.

However, executing simulated attacks generally increases the awareness of the employees and their involvement in security practices. Unfortunately, security management is not a silver bullet against security threats. At some point, these controls may fail as security incidents will happen. Security management processes also include incident management procedures to ensure that the organization can keep doing business and the incident impact is kept to a minimum. Security management allows us to use security technologies effectively.

It provides us with the tools to optimize the way people interact with technology so the risks that arise from these interactions are mitigated. Security management processes help us ensure that confidentiality, integrity, and availability of organizational assets are met. It also serves to ensure regulatory compliance and recovery from security incidents. Implementing security management processes won't save us, unfortunately, from security incidents. But it will, for sure, reduce them and help organizations achieve their security goals.

1.5 THE CIA AND DAD TRIADS

The concepts of security principles, also known as the CIA and DAD triads, will jump out over and over. Each domain seems to go back into this concept and explain itself by the security principles. How the CIA triad and DAD triad are mutually exclusive.

The CIA triad

The CIA triad, officially known as the principles of security, consists of confidentiality, integrity, and availability.

- Confidentiality: This means that data is protected from unauthorized disclosure. Think about it this way: for Frodo to reach Mount Doom in Mordor, he had to keep the ring secret. He provided confidentiality to the ring until he reached Mount Doom to destroy it. The fellowship knew he had it, because they were authorized to know it was there, but no one else should know. (Frodo sucked at keeping that thing secret though.)
- Integrity: This means that data is protected from unauthorized manipulation. A message that has integrity was received exactly as it was intended with no unauthorized changes. This concept does not discern

between data at rest (saved data) or data in transport (when you send data). Instead, it is reliant on just the idea of prevention of unauthorized change of the data. When Frodo and the party received the letter from Gandalf in Bree, he had to break a seal. That seal was proof that the message was in fact untampered with and was from Gandalf. Albeit, a wax seal is not a very good method of verifying integrity nowadays.
- Availability: This means that data is available when needed by the intended user. This might be the easiest concept to grasp. Frodo had the protection of the mithril shirt available to him during his travels, even though at times it was unknown to his comrades. When he needed it (a pointy spear anyone?), it was available to him.

The DAD triad

Like every concept in security, the CIA triad can be a double-edged sword. Where there is a good side, there is an opposite bad side to consider as well. In the lack of each of the CIA triad, you are given the DAD triad.

- Disclosure: This is the opposite of confidentiality. An example of this is when Frodo let the inhabitants of the inn know he had the ring by accidentally putting it on, alerting Strider and Sauron in Mordor that he had the ring.
- Alteration: This is the opposite of integrity. If, for instance, an agent of Sauron had intercepted the letter Gandalf had sent Frodo and modified it to tell Frodo that Gandalf was going to fix everything on his own, and to stay in the Shire at all costs, then resealed it as if it was from Gandalf, we would have a perfect example of alteration.
- Denial: This is the opposite of availability. Frodo was denied his letter in the Shire, and the result was almost deadly for him.

How are the CIA and DAD triads mutually exclusive?

Each point of the CIA and DAD triangle are exact opposite of each other. If a CIA principle is absent, then a DAD principle is present. Thus, you cannot have both at the same time. You could not have both a denial and availability at the exact same time; it is either one or the other.

How can you relate the CIA triad in your everyday life?

These are very broad concepts, and as such, it is very easy to relate them to your daily activities. When you make a decision for which way you want to take to get to work, try to apply them:

- Confidentiality: I don't want any bad guys knowing my route to work, so I won't broadcast my entire route on Facebook.

- Integrity: I have to make sure my route isn't tampered with, so I don't make a wrong turn and am late for work.
- Availability: If I lose my 4G access, I want to make sure I can still access my directions, so I will download the directions to my phone.

In summary, these are very basic concepts to grasp. Each principle covers such a broad generalization that it is easy to place almost anything in relation to it. This becomes very important when you are attempting to study how each of the CISSP domains interrelates. So, when you are learning concepts throughout your studying, think, "Which of the three principles does this support, and how?" Because that is the kind of understanding you should be looking for.

1.6 ISMS PURPOSE AND OBJECTIVES

Introduction to information security policies

Information security policy (ISP) is a set of rules enacted by an organization to ensure that all users or networks of the IT structure within the organization's domain abide by the prescriptions regarding the security of data stored digitally within the boundaries the organization stretches its authority.

An ISP is governing the protection of information, which is one of the many assets a corporation needs to protect. The present writing will discuss some of the most important aspects a person should take into account when developing an ISP. Putting to work the logical arguments of rationalization, one could say that a policy can be as broad as the creators want it to be: basically, everything from A to Z in terms of IT security, and even more. For that reason, the emphasis here is placed on a few key elements, but you should make a mental note of the liberty of thought organizations have when they forge their own guidelines.

Elements of information security policy

Purpose: Institutions create ISPs for a variety of reasons:

- To establish a general approach to information security.
- To detect and forestall the compromise of information security such as misuse of data, networks, computer systems and applications.
- To protect the reputation of the company with respect to its ethical and legal responsibilities.
- To observe the rights of the customers; providing effective mechanisms for responding to complaints and queries concerning real or perceived non-compliance with the policy is one way to achieve this objective.

Scope (objective)

ISP should address all data, programs, systems, facilities, other tech infrastructure, users of technology, and third parties in a given organization, without exception.

The four components of security documentation are policies, standards, procedures, and guidelines. Together, these form the complete definition of a mature security program. The capability maturity model (CMM), which measures how robust and repeatable a business process is, is often applied to security programs. The CMM relies heavily on documentation for defining repeatable, optimized processes.

As such, any security program considered mature by CMM standards needs to have well-defined (objectives) policies, procedures, standards, and guidelines.

- Policy is a high-level statement of requirements. A security policy is the primary way in which management's expectations for security are provided to the builders, installers, maintainers, and users of an organization's information systems.
- Standards specify how to configure devices, how to install and configure software, and how to use computer systems and other organizational assets, to be compliant with the intentions of the policy.
- Procedures specify the step-by-step instructions to perform various tasks in accordance with policies and standards.
- Guidelines are advice about how to achieve the goals of the security policy, but they are suggestions, not rules. They are an important communication tool to let people know how to follow the policy's guidance. They convey best practices for using technology systems or behaving according to management's preferences.

Security policies

A security policy is the essential foundation for an effective and comprehensive security program. A good security policy should be a high-level, brief, formalized statement of the security practices that management expects employees and other stakeholders to follow. A security policy should be concise and easy to understand so that everyone can follow the guidance set forth in it. In its basic form, a security policy is a document that describes an organization's security requirements. A security policy specifies what should be done, not how; nor does it specify technologies or specific solutions. The security policy defines a specific set of intentions and conditions that will help protect an organization's assets and its ability to conduct business. It is important to plan an approach to policy development that is consistent, repeatable, and straightforward.

A top-down approach to security policy development provides the security practitioner with a roadmap for successful, consistent policy production.

Information Security and Management System 15

Figure 1.5 Security policy development process.

The policy developer must take the time to understand the organization's regulatory landscape, business objectives, and risk management concerns, including the corporation's general policy statements. As a precursor to policy development, a requirement mapping effort may be required in order to incorporate industry-specific regulation. We covered several of the various regulations as well as best-practice frameworks that security policy developers may need to incorporate into their policies. Figure 1.5 depicts the security policy development process.

A security policy lays down specific expectations for management, technical staff, and employees. A clear and well-documented security policy will determine what action an organization takes when a security violation is encountered. In the absence of clear policy, organizations put themselves at risk and often flounder in responding to a violation.

- For managers, a security policy identifies the expectations of senior management about roles, responsibilities, and actions that should be taken by management with regard to security controls.
- For technical staff, a security policy clarifies which security controls should be used on the network, in the physical facilities, and on computer systems.
- For all employees, a security policy describes how they should conduct themselves when using the computer systems, e-mail, phones, and voice mail.

A security policy is effectively a contract between the business and the users of its information systems. A common approach to ensuring that all parties are aware of the organization's security policy is to require employees to sign an acknowledgment document. Human Resources should keep a copy of the security policy documentation file in a place where every employee can easily find it.

Security policy development

When developing a security policy for the first time, one useful approach is to focus on the why, who, where, and what during the policy development process:

- Why should the policy address these particular concerns? (purpose)
- Who should the policy address? (responsibilities)
- Where should the policy be applied? (scope)
- What should the policy contain? (content)

For each of these components of security policy development, a phased approach is used.

Phased approach

If you approach security policy development in the following phases, depicted in Figure, the work will be more manageable:

- Requirements gathering which comprises regulatory requirements (industry-specific), advisory requirements (best practices), and informative requirements (organization-specific).
- Project definition and proposal based on requirements.
- Policy development.
- Review and approval.
- Publication and distribution.
- Ongoing maintenance (and revision).

After the security policy is approved, standards and procedures must be developed in order to ensure a smooth implementation. This will require the policy developer to work closely with the technical staff to develop standards and procedures relating to computers, applications, and networks.

Security policy contributors

Security policy should not be developed in a vacuum. A good security policy forms the core of a comprehensive security awareness program for employees, and its development shouldn't be the sole responsibility of the IT department. Every department that has a stake in the security policy should be involved in its development, not only because this enables them to tailor

Information Security and Management System 17

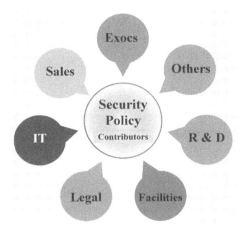

Figure 1.6 Example of security policy contributors.

the policy to their requirements but also because they will be responsible for enforcing and communicating the policies related to each of their specialties. Different groups and individuals should participate and be represented in order to ensure that everyone is on board, that all are willing to comply, and that the best interests of the entire organization are represented. When creating a security policy, the following groups may be represented. Figure 1.6 shows groups of security policy.

- Human Resources: The enforcement of the security policy, when it involves employee rewards and punishments, is usually the responsibility of the Human Resource (HR) department. HR implements discipline up to and including termination when the organization's policies are violated. HR also obtains a signature from each employee certifying that they have read and understood the policies of the organization, so there is no question of responsibility when employees don't comply with the policy.
- Legal: Often, an organization that has an internal legal department or outside legal representation will want to have those attorneys review and clarify legal points in the document and advise on particular points of appropriateness and applicability, both in the organization's home country and overseas. All organizations are advised to have some form of legal review and advice on their policies when those policies are applied to individual employees.
- Information technology: Security policy tends to focus on computer systems and specifically on the security controls that are built into the computing infrastructure. IT employees are generally the largest consumers of policy information.

- Physical security: Security (or facilities) departments usually implement the physical security controls specified in the security policy. In some cases, the IT department may manage the information systems' components of physical security.

Security policy audience

The intended audience for the security policies is all the individuals who handle the organization's information, such as:

- Employees.
- Contractors and temporary workers.
- Consultants, system integrators, and service providers.
- Business partners and third-party vendors.
- Employees of subsidiaries and affiliates.
- Customers who use the organization's information resources.

Figure 1.7 shows a representation of some example security policy audience members. Technology-related security policies generally apply to information resources, including software, web browsers, e-mail, computer systems, workstations, PCs, servers, mobile devices, entities connected on the network, software, data, telephones, voice mail, fax machines, and any other information resources that could be considered valuable to the business.

Organizations may also need to implement security policies contractually with business partners and vendors. They may also need to release a security policy statement to customers.

Figure 1.7 Security policy audience.

Policy categories

Security policies can be subdivided into three primary categories:

Regulatory: For audit and compliance purposes, it is useful to include this specific category. The policy is generally populated with a series of legal statements detailing what is required and why it is required. The results of a regulatory requirements assessment can be incorporated into this type of policy.

Advisory: This policy type advises all affected parties of business-specific policy and may include policies related to computer systems and networks, personnel, and physical security. This type of policy is generally based on the best practices of security.

Informative: This type of policy exists as a catch-all to ensure that policies not covered under regulatory and advisory are accounted for. These policies may apply to specific business units, business partners, vendors, and customers who use the organization's information systems.

The security policy should be concise and easy to read, in order to be effective. An incomprehensible or overly complex policy risks being ignored by its audience and left to gather dust on a shelf, failing to influence current operational efforts. It should be a series of simple, direct statements of senior management's intentions.

The form and organization of security policies can be reflected in an outline format with the following components:

Author: The policy writer.
Sponsor: The executive champion.
Authorizer: The executive signer with ultimate authority.
Effective date: When the policy is effective; generally, when authorized.
Review date: Subject to agreement by all parties; annually at least.
Purpose: Why the policy exists; regulatory, advisory, or informative.
Scope: Who the policy affects and where the policy is applied.
Policy: What the policy is about.
Exceptions: Who or what is not covered by the policy.
Enforcement: How the policy will be enforced, and the consequences for not following it.
Definitions: Terms the reader may need to know.
References: Links to other related policies and corporate documents.

1.7 FRAMEWORKS

The topics included in a security policy vary from organization to organization according to regulatory and business requirements. We refer to these topics together as a framework. Organizations may prefer to take a control objective–based approach to creating a security policy framework. For instance, government agencies may take a Federal Information Security Management Act (FISMA)–based approach. The Federal Information Security Management Act of 2002 imposes a mandatory set of processes that must follow a combination of Federal Information Processing Standards (FIPS) documents, the NIST Special Publications 800 series, and other legislation pertinent to federal information systems.

Policy categories

NIST Special Publication 800-53, Recommended Security Controls for Federal Information Systems and Organizations, control objectives are organized into 18 major categories. Control objective subsets exist for each major control category and equal at least 170 control objectives. NIST SP 800-53 is a good starting point for any organization interested in making sure that all the basic control objectives are met regardless of the industry and whether it is regulated.

Additional regulations and frameworks

An organization that must comply with HIPAA (described in Chapter 3) may map NIST SP 800-53 control objectives to the HIPAA Security Rule. HIPAA categorizes security controls (referred to as safeguards) into three major categories: administrative, physical, and technical. As an example, CFR Part 164.312 section (c)(1), which requires protection against improper alteration or destruction of data, is a HIPAA-required control that maps to NIST 800-53 system and information integrity controls. Some organizations may wish to select a framework based on COBIT (Control Objectives for Information and Related Technology). COBIT is an IT governance framework and supporting toolset that allows managers to bridge the gap between control requirements, technical issues, and business risks. Developing policy from a COBIT framework may take considerable collaboration with the Finance and Audit departments. Other organizations may need to combine COBIT with IT Infrastructure Library (ITIL) to ensure that service management objectives are met. ITIL is a cohesive best-practices framework drawn from the public and private sectors internationally. It describes the organization of IT resources to deliver business value and documents processes, functions, and roles in IT service management. Still other organizations may wish to follow the OCTAVE (Operationally Critical Threat, Asset, and Vulnerability Evaluation) framework. OCTAVE is a risk-based strategic assessment and planning technique for security from CERT (Carnegie Mellon University). And yet others may need to incorporate the International Standards Organization (ISO) Family (27001 and 27002) from the International Standards Organization. ISO is a framework of standards that provides best practices for information security management. Depending on which regulated industry an organization finds itself in, it is important to take the time to select an appropriate framework and to map out the regulatory and business requirements in the first phase of development.

Security management policies

Managers have responsibilities for security just as employees do. Detailing expectations for managers is crucial to ensure compliance with senior management's expectations.

- Employee nondisclosure agreements: All employees must sign a nondisclosure agreement that specifies the types of information they are prohibited from revealing outside the organization. The agreement must be signed before the employee is allowed to handle any private information belonging to the organization. Employees must be made aware of the consequences of violating the agreement, and signing the agreement must be a condition of employment such that the organization may not employ anyone who fails to sign the agreement.
- Nondisclosure agreements: All business partners wishing to do business with the organization must sign a nondisclosure agreement that specifies the types of information they are prohibited from revealing outside the organization. The agreement must be signed before the business partner is allowed to view, copy, or handle any private information belonging to the organization.
- System activity monitoring: All internal information system servers must be constantly monitored, $24 \times 7 \times 365$, by trained security analysts. At least the following activities must be monitored:
  - Unauthorized access attempts.
  - Root or administrator account usage.
  - Nonstandard behavior of services.
  - Addition of modems and peripherals to systems.
  - Any other relevant security events.
- Software installation monitoring: All software installed on all servers and end-user systems must be inventoried periodically. The inventory must contain the following information:
  - The name of each software package installed on each system.
  - The software version.
  - The licensing status.
- System vulnerability scanning: All servers and end-user systems must be periodically scanned for known vulnerabilities. The vulnerability scan must identify the following:
  - Services and applications running on the system that could be exploited to compromise security.

- File permissions that could grant unauthorized access to files.
- Weak passwords that could be easily guessed by people or software.

Security document lifecycle: All security documents, including the corporate security policy, must be regularly updated and changed as necessary to keep up with changes in the infrastructure and in the industry.

Security audits: Periodic security audits must be performed to compare existing practices against the security policy.

Penetration testing: Penetration testing must be performed on a regular basis to test the effectiveness of information system security.

Security drills: Regular "fire drills" (simulated security breaches, without advance warning) must take place to test the effectiveness of security measures.

Extranet connection approval: All extranet connections require management approval before implementation.

Non-employee access to corporate information: Non-employees (such as spouses) are not allowed to access the organization's information resources.

New employee access approval: Manager approval is required for new employee access requests.

Employee access change approval: Manager approval is required for employee access change requests.

Contractor access approval: Manager approval is required for contractor access requests.

Employee responsibilities: The following categories of responsibilities are defined for corporate employees. These categories consist of groupings of responsibilities that require differing levels of access to computer systems and networks.

They are used to limit access to computers and networks based on job requirements and to implement the principles of least privilege and separation of duties.
- General user.
- Operator.
- System administrator.
- Customer support staff.
- Customer engineer.
- Management.

Security personnel responsibilities: The following categories of responsibilities are defined for security personnel. These categories consist of groupings of responsibilities within the security organization that require differing levels of access to security information and systems based on job function, in order to implement the principles of least privilege and separation of duties.
- Security architect.
- Facility security officer.

- Security manager.
- Technical security administrator.

Employee responsibility for security: All corporate employees are responsible for the security of the computer systems they use and the physical environment around them.

Sensitive HR information: Sensitive HR information (such as salaries and employee records) must be separated and protected from the rest of the corporate network.

Security policy enforcement of this corporate security policy is the responsibility of the corporate Human Resources department.

HR new hire reporting: HR must report required information about new hires to system administrators 1 week in advance of the new employee's start date.

HR termination reporting: HR must report required information about terminations to system administrators 1 week before the termination date, if possible, and no later than the day of termination.

Contractor information reporting: HR is responsible for managing contractor information and providing this information to system administrators.

Background checks: HR must perform background checks on new employee applicants.

Reference checks: HR must perform reference checks on new employee applicants.

1.8 SECURITY STANDARDS

A standard is somewhat more detailed than a policy. Standards describe how to comply with the policy, and because they are associated with policies, they should be considered mandatory. Standards are the extension of the policy into the real world – they specify technology settings, platforms, or behaviors. Security managers responsible for IT infrastructure will usually spend more time writing standards than they spend on the policy.

Much of the information contained in Chapters 21 and 22 of this book pertains to settings for Unix and Windows systems. Those settings would typically be the level of detail that is included in standards. Compare the information in those chapters against the set of policy statements listed in the previous section of this chapter. You'll see that policy statements are simple, direct, and somewhat general. Standards interpret the policy to the level of specifics needed by a subject matter expert.

Security standard example

The following is a sample of a security standard. This is part of a standard for securing Linux servers. It is intended to establish a baseline set of configurations that would establish common settings across all Linux platforms on the network. Notice that the level of detail is very deep – only an experienced system administrator would be able to understand some of these instructions. That is typical of a standard, as opposed to a policy, which everyone should be able to understand regardless of their level of expertise.

- Purpose: The purpose of this standard is to define the software and hardware configurations required to secure Linux servers. It defines security settings for operating systems and software that are required by policy.
- Scope: This standard is to be used by system administrators responsible for the administration of computers using the Red Hat Enterprise Linux operating system.
- Responsibilities: The security manager is responsible for defining this standard 3.2. The server team is responsible for following this standard.

1.9 STANDARD

Services

Specific services that are required for the general operation of the systems and resident vendor applications services are to be reviewed for security risks and approved by the security manager. Services that are not needed are to be disabled during boot.

Initial password and login settings

- All accounts for system administrators are to be added as local accounts in the /etc/passwd and /etc/shadow files. NIS is not to be used for password verification.
- Privileged user accounts require IT system operations and applications manager approval before being placed on the system.
- No developer accounts are allowed on production servers.
- All administration user accounts are to be set with 90-day password aging, 7-day notification of password expiration, and 7-day password minimum.
- All root and application administrator accounts are to be reviewed and will have a scheduled password change by operations administrators once every 90 days.
- The default login setting is to be set to lock out the session after three failed password login attempts.
- Default password settings must enforce a minimum of eight characters.
- The ability to log in directly over the network to the root account must be disabled.

Send mail

The send mail service is to be disabled on all non-mail servers unless required by an application running on the system. Applications requiring send mail services must first be approved by the IT system operations manager.

> Banner/notice: Configure the login banner with the standard warning notice.
> Logging:
> - Turn on logging for Internet standard services.
> - Turn on logging for LOG_AUTHPRIV facility.
> - Log connection tracing to inetd/xinetd and messages sent to AUTH facility.
> - Send all kernel authorization, debug, and daemon notices to a syslog server for monitoring, reviewing, and archiving.

1.10 SECURITY PROCEDURES

Procedures are step-by-step instructions to perform a specific task.

Security procedure example

In this example, notice that the level of detail is more specific than that found in both policies and standards. The procedure is a set of instructions that a system administrator would perform when sitting at the keyboard of the computer being built. Most people will not understand this information – it is very specialized and intended only for someone who is a system administrator. The type of specialized information found in a security procedure is usually very job-specific.

- Purpose: This procedure is intended for the security installation of Apache web servers. It defines the steps necessary to ensure a secure installation that complies with security policy.
- Scope: This procedure is to be used by system administrators responsible for installing the Apache HTTP server.
- Responsibilities:
  - The security manager is responsible for defining this procedure.
  - Any system administrator installing Apache HTTP server on the network is responsible for following this procedure.

Apache web server security procedure

Compile and install the server software as follows:

- ./configure   --prefix=/usr/local/apache   --disable-module=all   --serveruid=apache   --server-gid=apache   --enable-module=access

--enable- module=log_ config --enable-module=dir --enable-module=mime --enable- module=auth
- make
- su
- umask 022
- make install
- chown -R root:sys /usr/local/apache

The next step is to limit Apache processes' access to the file systems. Start this process by creating a new root directory structure under the /chroot/httpd directory:

- mkdir -p /chroot/httpd/dev
- mkdir -p /chroot/httpd/etc
- mkdir -p /chroot/httpd/var/run
- mkdir -p /chroot/httpd/usr/lib
- mkdir -p /chroot/httpd/usr/libexec
- mkdir -p /chroot/httpd/usr/local/apache/bin
- mkdir -p /chroot/httpd/usr/local/apache/logs
- mkdir -p /chroot/httpd/usr/local/apache/conf
- mkdir -p /chroot/httpd/www

Next, create the special device file: /dev/null:

- ls -al /dev/null
- crw-rw-rw- 1 root wheel 2, 2 Mar 14 12:53 /dev/null
- mknod /chroot/httpd/dev/null c 2 2
- chown root:sys /chroot/httpd/dev/null
- chmod 666 /chroot/httpd/dev/null

Add the following line to the /etc/rc.conf file:

- syslogd_flags="-l /chroot/httpd/dev/log"

Restart the system.

Copy the main httpd program into the new directory tree with all necessary binaries and libraries, as follows:

- localhost# ldd /usr/local/apache/bin/httpd

Copy the files to the new root directory structure:

- cp /usr/local/apache/bin/httpd /chroot/httpd/usr/local/apache/ bin/
- cp /var/run/ld-elf.so.hints /chroot/httpd/var/run/

- *cp /usr/lib/libcrypt.so.2 /chroot/httpd/usr/lib/*
- *cp /usr/lib/libc.so.4 /chroot/httpd/usr/lib/*
- *cp /usr/libexec/ld-elf.so.1 /chroot/httpd/usr/libexec/*

1.11 SECURITY GUIDELINES

Guidelines give advice. They are not mandatory – they are just suggestions on how to follow the policy. Guidelines are meant to make life easier for the end user, as well as for the security manager who wrote the policy because they help people understand how to meet the goals set by the security policy.

Security guideline example

In this example, the password complexity rules of the password policy are translated into a set of easy-to-follow suggestions. There may be other ways to select a password to be compliant with the policy, but these guidelines are intended to simplify the process for the end users while at the same time allowing them to make strong passwords. Notice that unlike standards and procedures, the material is easy for everyone to read and understand.

- Purpose: These guidelines are meant to give you some ideas about how to create a good password. Our password policy requires a certain amount of complexity, which can result in difficult-to-remember passwords, but these guidelines should help you comply with our password policy while at the same time making it easier for you to choose a memorable password.
- Scope: These guidelines are for all people who have computer accounts on our network.
- Responsibilities: The security manager is responsible for defining, maintaining, and publishing these guidelines.

Password selection guidelines

Do

- Use as many different characters as possible including numbers, punctuation characters, and mixed upper- and lowercase letters. Choosing characters from the largest possible range will make your password more secure.
- Use both upper- and lowercase letters.
- Use at least one number and one punctuation mark.
- Select passwords that are easy to remember, so they do not have to be written down.

Don't

- Your name, the names of any family or friends, names of fictional characters.
- Phone number, license, or social security numbers.
- Any date.
- Any word in the dictionary.
- Passwords of all the same letter or any variation on the word "password".
- Simple patterns on the keyboard, like qwerty.
- Any word spelled backward.

Suggestions

- Use the first one or two letters of each word in a phrase, song, or poem you can easily remember. Add a punctuation mark and a number.
- Or, use intentionally misspelled words with a number or punctuation mark in the middle.
- You can also alternate between one consonant and one or two vowels and include a number and a punctuation mark. This provides a pronounceable nonsense word that you can remember.
- Or, you can choose two short words and concatenate them together with a punctuation character between them.
- Or, interlace two words or a word and a number (like a year) by alternating characters.

1.12 COMPLIANCE VS. CONFORMANCE

Compliance

Compliance has been commonly defined as doing what you are told to do, and in simple terms, this does prove to be correct; however, when it applies to our industry and the overarching quality management realm, its definition is far more complex. One of the key traits to compliance is the fundamental basis of legality, a standard, or requirement in which legal channels may pursue their enforcement or application through jurisdictional channels or authorities. This legal foundation of course implies the second key trait of compliance, which is that it is a requirement or expectation that is externally applied upon a person(s) or organization. The complexity lies in who has the authority or right to impose such a legal expectation, a category that can encompass a broad range of situations and entities.

A common example of an expectation of compliance comes through the widely applied Occupational Health and Safety Standards that mandate the treatment of workers and the rights and responsibilities of employees, employers, and even the general public in relation to safety at the workplace. This standard is enforced by Occupational Health and Safety officers employed by the government that authors the standard, an entity that has the capability to enforce the legal obligations to this regulation upon any person(s) or company that is found to be in non-compliance with the regulation.

Conformance

Conformance on the other hand is seen as the voluntary adherence to doing something in a recognized way. Conformance is widely applied in every facet of life in and out of the workplace, cultural norms and social expectations of how to act are a widely applied form of conformance, and the exercise of rebelling against or challenging these is at its most basic form, a non-conformance. The spirit of conformity when it applies to quality systems and their application comes from the desire to improve and better the processes and procedures that a company may have in place to ensure consistent results when providing products or services to a client.

Conformance comes from within, or from the bottom up, we choose to remove our shoes when entering someone's home as a form of voluntary conformance to a widely recognized cultural standard, this standard may change in different situations or locations based on what is recognized as acceptable in that area and isn't imposed upon us by any external influence. To get everything in order, in alignment and cohesion and to create a status quo as an expectation for the operation of a company or application of a standard or process is the heart of conformity in our industry.

Special applications

Why is there confusion surrounding this topic? When I chose to research the difference between the two terms, I found a myriad of conflicting reports and viewpoints on the subject, arguments and disagreements depending on the perspective of the author or the application of the term. In fact, in many cases sources would say that the terms are downright interchangeable and that the disagreement in the usage of the term actually stems from the difference between standard "British English" and standard "American English" as so many terms or spellings fall to. Through investigation into the origins of the words and their applications across a wide berth of situations, I was able to differentiate the difference and come to a conclusion because of one distinct caveat. "If someone mandates you meet the requirements of a standard or test method then conformance becomes compliance (i.e., your conformity is required in order for you to comply)."

This very important distinction and how it applies to the codes and standards that I am familiar with provide a clear difference that is easily demonstrated and repeated. For example, ISO:9001 is a commonly applied international standard for the structure and application of a quality management program. This standard is a matter of conformance when viewed in solidarity, a widely recognized way of doing something that a company or organization will choose to adhere to in order to better their own processed and procedures, and in fact improve the company as a whole.

However, when we take this standard and apply it in conjunction with a matter of compliance such as a client contract, the conformance standard becomes a matter of compliance in itself. If we do not conform to the standard, we will not comply with the contract and could be penalized by law, an external force is implying this voluntary standard, making it an involuntary matter of compliance. Many companies deal with repeating similar situations within their industry in which contractual requirements are the norm, a standard such as ISO 9001 will be continually stipulated in contractual legal agreements making ISO 9001, in their eyes, a matter of compliance. Many other organizations choose to adhere to the standard as a widely recognized guideline, making it a matter of conformance.

Conclusion on compliance and conformance

In summary, the main difference between compliance and conformance is the source of the implementation of whichever guideline or standard is in question. Externally applied with legal ramifications is a matter of compliance, internally applied with voluntary adherence is conformance. But don't forget that some standards of compliance can be conformed too, and some guidelines for conformance can become a matter of compliance.

BIBLIOGRAPHY

1. Joshi, C. and Singh, U.K., 2017. Information security risks management framework–A step towards mitigating security risks in university network. *Journal of Information Security and Applications*, 35, pp.128–137.
2. Chopra, A. and Chaudhary, M., 2020. *Implementing an information security management system*. Apress.
3. Fomin, V.V., Vries, H. and Barlette, Y., 2008, September. ISO/IEC 27001 information systems security management standard: Exploring the reasons for low adoption. In Euromot 2008 Conference, Nice, France.
4. Padmavathi, D.G. and Shanmugapriya, M., 2009. A survey of attacks, security mechanisms and challenges in wireless sensor networks. arXiv preprint arXiv:0909.0576.
5. Cohen, F., 1997. Information system attacks: A preliminary classification scheme. *Computers & Security*, 16(1), pp.29–46.
6. Visintine, V., 2003. An introduction to information risk assessment. *SANS Institute*, 8, p.116.
7. Hoff, K. and Stiglitz, J.E., 1990. Introduction: Imperfect information and rural credit markets: Puzzles and policy perspectives. *The World Bank Economic Review*, 4(3), pp.235–250.

8. Rainer, R.K. and Prince, B., 2021. *Introduction to information systems*. John Wiley & Sons.
9. Tsohou, A., Karyda, M., Kokolakis, S. and Kiountouzis, E., 2015. Managing the introduction of information security awareness programmes in organisations. *European Journal of Information Systems*, 24(1), pp.38–58.
10. Ford, N., 2015. *Introduction to information behaviour*. Facet Publishing.
11. Fisher, C., Lauría, E. and Chengalur-Smith, S., 2012. *Introduction to information quality*. AuthorHouse.
12. Giri, D., Barua, P., Srivastava, P.D. and Jana, B., 2010, June. A cryptosystem for encryption and decryption of long confidential messages. In International Conference on Information Security and Assurance (pp. 86–96). Springer, Berlin, Heidelberg.

Chapter 2

Audit Planning and Preparation

INTRODUCTION

The board of directors of various sectors acts as trustees and ought to secure the cooperative's financial records and annual audit. The board is liable in case of any negligence, and it is always advisable to maintain a complete double-entry booking system, entire monthly financial statements, and annual audit with supporting documents. The trustee must ensure zero misstatements of facts, accurate and consistent accounting methodology, implementation of generally accepted accounting principles (GAAP), and complete disclosure. The auditor deputed by the trustee should plan to complete the work proficiently and within time by following accounting policies, systems, and procedures. An auditor must be honest, sincere, confidential by nature and trained, experienced, and competent professionals, and should be prone to new legal developments in rules and regulations of accounting and auditing.

Audit competence and evaluation methods assist to perform better in auditing. Reporting on the financial statements by satisfying all the criteria plays a crucial role in auditing. The law of conservation needs the sum of the inputs to be equal to the sum of the outputs, and this applies and maps to the Process Principles. PEEMMM with case studies helps us understand the importance of inputs and outputs through process equilibrium. ISO 27001 Audit Checklist standardizes the security of proper financial management and auditing. Various business companies, banks, government, and private sectors follow PEEMMM for better understanding and visualization. ISO 27001 Checklist is prepared by an expert panel of IRCA Principal Auditors and Lead Instructors of Information Security Management System.

2.1 REASONS FOR AUDITING

Securing an annual audit of the cooperative's financial records is the responsibility of the board of directors. Because the board acts as the trustee of the cooperative's assets, it is responsible for safeguarding, auditing, and appraising the cooperative's financial resources. The audit is a fundamental part of this trustee responsibility, and the cost of the audit should be considered a normal business expense.

The fiduciary responsibilities of the board should not be taken lightly. If a board of directors is negligent in establishing and monitoring the operations of their cooperative, directors could be held liable. The first step in fulfilling this obligation is a complete, double-entry bookkeeping system; the second is monthly financial statements; and the third is an annual audit of the accounting records and supporting documents.

The fiduciary/trustee responsibility of the board of directors translates into five specific reasons why the board must provide for an annual audit of the cooperative's accounting records:

1. Prevent deliberate misstatement of fact. Misstatement of fact may occur for many reasons, such as to hide poor decisions or to cover fraud.
2. Ensure the judgment decisions are not unduly biased in favor of management. It is the board's duty to develop and implement the accounting system and the management's duty to maintain the books.
3. Ensure records are dependable. Accounting methods should be accurate as well as consistent. An audit will identify shortcomings in accuracy and/or consistency. Procedures lacking consistency fail to be dependable for purposes of analysis and decision making.
4. Ensure generally accepted accounting principles (GAAPs) have been consistently followed. The American Institute of Certified Public Accountants (AICPA) has set guidelines, laws, or rules in accounting practices to prepare financial statements. This will allow comparisons with organizations whose audits also follow standard practices.
5. Ensure that the disclosure is complete. In many cases, what is not reported is often more important than what is reported. An audit will help the board of directors ensure that full disclosure of the financial wellbeing of the cooperative business has been made.

2.2 AUDIT PRINCIPLES

2.2.1 Planning

An auditor should plan to complete his work efficiently and well within time. To plan work accordingly, an auditor handles the following:

- Accounting system and policies.
- Internal control system of organization.
- Determination of audit procedures and coordinating audit work.

2.2.2 Honesty

An auditor must have an impartial attitude and should be free from any interest. He should be honest and sincere in his work, and he should do his work without any bias and prejudice.

2.2.3 Secrecy

An auditor should keep confidential all the information acquired by him during his audit. He should not share the information with anyone without the permission of the client and that too the information can be shared with the client's permission only when it is bound to be so.

2.2.4 Audit evidence

An auditor should adhere to substantive and compliance procedures for collecting audit evidences before conducting an audit. Through substantive procedures, an auditor may collect evidences regarding the accuracy, completeness, and validity of data, and through compliance procedures, he may collect evidences regarding the internal control system as used in the client's organization.

2.2.5 Internal control system

It is the primary responsibility of a company to keep an adequate internal control system in the organization. On the basis of such internal control system, an auditor can determine the nature, timing, and audit procedure to be applied to conduct his audit.

2.2.6 Skill and competence

Audit should be done by trained, experienced, and competent persons, and audit staff should be updated with all the developments in accounting, auditing, and legal rules and regulations as amended from time to time.

2.2.7 Work done by others

An auditor is permitted to rely on work done by others, but he should exercise due diligence when referring to it. He should mention the source of reference thereof in his report.

2.2.8 Working papers

An auditor should prepare and preserve all the necessary documents as obtained during his audit. These documents can be used by him as audit evidences.

2.2.9 Legal framework

All business activities should be run adhering to the rules and regulations stipulated in the legal framework. This is to safeguard the interests and rights of the interested parties.

2.2.10 Audit report

On the basis of the review and assessment of the audit evidences, an auditor should express his opinion regarding the financial statements of an organization:

- Financial statements are prepared using acceptable accounting principles.
- Financial statements should comply with all relevant statutory requirements.
- All material matters are disclosed, and proper presentation of financial statements is done subject to statutory requirements.

2.3 PROCESS OF AUDIT PROGRAM MANAGEMENT

The Case Western Reserve's Board of Trustees and management place assets at risk to achieve established priorities and goals. A key function of the Office of Internal Audit Services is to understand, audit, and report to management and the Board of Trustees how that risk is being managed. Knowing what areas to audit and where to commit resources is an integral part of managing the internal audit function.

To identify areas of potential risk, each year, the Office of Internal Audit Services performs a thorough risk assessment of all university management centers, operating units, and significant departments. From this assessment, an audit plan is developed and presented to the audit committee for approval.

The plan addresses high-risk areas as well as allocates time for special ad hoc projects. In intervening years, the risk assessment is updated through data analysis and interviews with senior executives across the university. If necessary, the audit plan is adjusted for any changes to the university's risk assessment.

We believe that the university is best served if the audit plan is a dynamic document that continually adjusts to changes in the environment. Therefore, if your management center or department has a need for our services, please contact us.

Depending on the relative risk associated with your need and the amount of time necessary to fulfill your request, the Office of Internal Audit Services will communicate what level of assistance we will be able to provide. At a minimum, we will be available to offer guidance and advice throughout any project you perform on your own.

In most cases, expect to receive a notification when you or your department is to be audited.

- Expect to understand the audit's purpose and objective.
- Expect to provide your ideas or concerns regarding the audit.
- Expect to be treated with respect and courtesy.
- Expect to provide documentation; some may be confidential.
- Expect confidential information to remain confidential.
- Expect to answer all questions honestly.
- Expect its release.

2.3.1 Preparing for an audit

- Have all requested materials/records ready when requested.
- Organize files so we minimize disruption of your day.
- Provide complete files.
- Please make yourself available during the time of the audit and communicate any planned absences.
- Provide workspace for auditors if requested.

2.3.2 Audit process

Step 1: Planning
The auditor will review prior audits in your area and professional literature. The auditor will also research applicable policies and statutes and prepare a basic audit program to follow.

Step 2: Notification
The Office of Internal Audit Services will notify the appropriate department or department personnel regarding the upcoming audit and its purpose, at which time an opening meeting will be scheduled.

Step 3: Opening meeting
This meeting will include management and any administrative personnel involved in the audit. The audit's purpose and objective will be discussed as well as the audit program. The audit program may be adjusted based on information obtained during this meeting.

Step 4: Fieldwork
This step includes the testing to be performed as well as interviews with appropriate department personnel.

Step 5: Report drafting
After the fieldwork is completed, a report is drafted. The report includes such areas as the objective and scope of the audit, relevant background, and the findings and recommendations for correction or improvement.

Step 6: Management response
A draft audit report will be submitted to the management of the audited area for their review and responses to the recommendations. Management responses should include their action plan for correction.

Step 7: Closing meeting

This meeting is held with department management. The audit report and management responses will be reviewed and discussed. This is the time for questions and clarifications. Results of other audit procedures not discussed in the final report will be communicated at this meeting.

Step 8: Final audit report distribution

After the closing meeting, the final audit report with management responses is distributed to department personnel involved in the audit, the President, Provost, and Chief Financial Officer, and CWRU's external accounting firm.

Step 9: Follow-up

Approximately 6 months after the audit report is issued, the Office of Internal Audit Services will perform a follow-up review. The purpose of this review is to conclude whether or not the corrective actions were implemented.

2.4 AUDIT COMPETENCE AND EVALUATION METHODS

In Table 2.1, we will learn the various types/classes of audit and their basis.
It lists out the different types of audit.
Let us now understand the important classifications of audit.

Table 2.1 Few basics of auditing to learn and the various types or classes of audit which makes a learner to apply the right basic with appropriate type

Basis	Type
Scope	**Specific audit** – Cash audit, cost audit, standard audit, tax audit, interim audit, audit in depth, management audit, operational audit, secretarial audit, partial audit, post & vouch audit, etc., are common types of specific audit. **General audit** – It can be an internal or an independent
Activities	• Commercial • Non-commercial
Organization	• Government • Private
Legal	**Statutory** – Insurance company, electricity company, banking companies, trust, company, corporations, cooperative societies. **Non-statutory** – Individual, firm, sole trader, etc.
Examination methods	• Periodicals • Continuous
Who conducts	• Internal audit • Independent audit

2.4.1 Audit of individuals

Sources of income of any individual may be from his investments, property, shares, commission as agent, interest income, etc.

Following are the purposes and benefits if anyone opts for an audit:

- To know the correct income from all of his sources.
- To assure accuracy.
- To prevent and detect any fraud or misappropriation.
- To be helpful and useful in income tax assessment.
- To keep a moral check on accountant and agent.

2.4.2 Audit of sole trader's books of accounts

The scope of audit will depend on the instructions and agreement between auditor and sole proprietor, the latter being an individual owner of the business; the sole proprietor decides the scope of audit.

The purpose and benefit of audit in a sole trader's business are almost the same as for an individual. Following are some additional benefits.

- Assurance about proper vouchers of his expenditure and preparation of his accounts with accuracy and correctness.
- Assurance about a true and fair picture of his business income and expenditure.
- His accounts can be compared with the previous years'.

2.4.3 Audit of partnership firm

An auditor for a partnership firm may be appointed by the partners with mutual consent. Mutual agreement between partners and auditor is based on the latter's rights, liabilities, and the scope of his audit. Reference to the partnership deed is a must for an auditor, and he should refer to the Partnership Act, 1932, in case where the partnership deed is silent. Certificate of an auditor will contain points related to the following.

- Reliability of accounts depending upon the nature of business.
- If any restrictions and limitations are imposed by the partners on his audit scope.
- Whether the auditor got all the required information and explanations or not.

Important provision of Partnership Act

An auditor should refer to the following provisions of the Partnership Act, 1932, where the partnership deed is silent.

- A minor can be admitted to a firm as a partner only for profits, he will not be liable for any loss.

- Property of the firm can be exclusively used by the partners for business purposes.
- Partners will share profit and loss equally.
- There is no entitlement of any remuneration or salary to any of the partners.
- Six percent interest on capital will be paid to the partner in case of any addition to capital made by any partner in excess of the agreed amount.
- Interest on capital will be payable out of profits only.
- Goodwill of the firm will be treated as assets of the firm at the time of dissolution of the firm.
- At the time of dissolution of the firm, the settlement of account will be done in the following order:
- Out of profit,
- Out of capital,
- By the partners individually in their profit-sharing ratio.

2.4.4 Government audit

The Government of India maintains a separate department known as the Accounts and Audit Department, and this department is headed by the Comptroller and Auditor General of India, which works only for government offices.

Important features of the government audit

- In almost every government department, prior sanction is a must before any payment of expenditure.
- Before making any payment, a preliminary examination of bills is done by the treasury officer.
- The nature of government audit is always continuous due to a large number of transactions and a huge amount of expenditure.
- A major portion of the accounts is prepared by the Accounts and Audit department which works independently.

Objectives

The following are the main objectives of government audit:

- To check and ensure that prescribed rules and regulations have been followed while making payments.
- To ensure that expenditure should not be excessive.
- To check and verify physical stock, stores, and spares along with their proper valuation. Stock-taking should be done at regular intervals,

and the recording of stock in the stock register should be done correctly and up-to-date.
- To check whether every payment is sanctioned by proper authority or not.
- To ensure that expenditure should be done in public interest only by the right person and should be paid to the right person.
- To ensure that no expenditure should be incurred for any personal benefit of any authority.
- To give suggestions for any kind of improvement in efficiency and economy.
- To verify that the amount due from others is properly recorded in the books and also to verify that such amount is regularly recovered.

2.4.5 Statutory audit

Where the appointment of a qualified auditor is compulsory as per the law is called a statutory audit. The following are the essential characteristics of statutory audit:

- An auditor must be a qualified accountant.
- Norms of the appointment of the auditor are provided by the law. The rights, duties, and liabilities of an auditor are as defined by the statute; the management cannot make any changes to it.
- Organizations cannot restrict the scope of statutory audit.
- Statutory audit provides a true and fair view of the financial position to shareholders and members of an organization. It helps the shareholders to keep themselves protected from any fraud and misrepresentation.
- Statutory audit is a compulsory audit. Auditor is an independent person and management doesn't have any control over his work.
- Following stakeholders are covered under the statutory or compulsory audit.

2.4.6 Audit of companies

First time in India, the Indian Companies Act, 1913, made it compulsory for joint stock companies to get their accounts audited by a qualified person (chartered accountant). Appointments, duties, qualifications, powers, and liabilities are amended through the Companies Act, 1956 and 2013.

2.4.7 Audit of trust

The Public Trust Act provides a compulsory audit of accounts by a qualified auditor. Conditions and terms as laid down in trust deed are the basis on which accounts of trusts are maintained. Any beneficiary of trust does not have control or access over accounts of trust; therefore, there are more chances of fraud and misappropriations.

2.4.8 Audit of cooperative societies

The Companies Act is not applicable to societies; cooperative societies are established under the Cooperative Societies Act, 1912. It is a must for a qualified accountant to have the required expertise, and he should be updated with various amendments to the Act. An auditor should also have knowledge of the by-laws of this act.

2.4.9 Audit of other institutions

Banks, insurance companies, and electricity companies are audited as per the provisions of Special Act of the Parliament.

Cost audit

Services of qualified cost accountants are necessary to have full control of the records of costs and cost variations. Big business houses and manufacturing units do understand the importance of cost accounting. Cost auditors check the work done by cost accountants to ensure the correctness of the accounting.

Objectives of cost audit

- To verify the arithmetical accuracy of cost accounting.
- To help management to take the decision about production and cost variations.
- To detect errors and frauds.
- To have control over the cost accounting department.
- To give suggestions about the efficiency of material, labor, and machine.

2.4.10 Tax audit

Under the provision of section 44AB of the Income Tax Act, 1961, every person carrying a business/profession is required to get his accounts audited, if the total turnover or gross receipts during the previous year exceed Rs. 100 lacs in case of business and Rs. 25 lacs in case of profession.

Profit and loss account of a business or profession is adjusted according to the provision of the Income Tax Act; therefore; accounting profit and tax profit differ. The reason behind the difference in profit or loss may be because of the following:

- Amount of depreciation.
- Under the Income Tax Act, certain expenses are allowed only on the basis of actual payment and those should be within the prescribed

time as provided by law, like the payment of provident fund, ESI, interest to financial institutions, VAT/Central Sales Tax, employee-related payments, etc.

2.4.11 Balance sheet audit

Balance sheet audit is very popular in the United States. It is an annual audit and covers each and every item of nominal accounts as appeared in the profit and loss account, assets, liabilities, reserves, provisions, stocks, and surplus. It is also done by highly skilled accountants.

Continuous audit

Under continuous audit, each and every transaction of the business is checked by the auditor regularly. A continuous audit is required in large organizations where the number of transactions is very high, the internal control system is not effective, periodicals statements are required, and final accounts are prepared immediately after the close of financial years like banks.

> **Advantages** – Complete checking of records, up-to-date accounts, moral-check on staff, and early finalization of financial statements are the main advantages of the continuous audit.
> **Disadvantages** – High cost of continuous audit, mechanical work of auditor, chances of unhealthy relations with staff due to frequent visits, etc., are main disadvantages of the continuous audit.

Annual audit

In an organization where the number of transactions is not large, an auditor usually comes after the close of the financial year and completes his audit work in continuous sessions. In the case of small business houses, an annual audit gives satisfactory results.

> **Advantages** – The work that is done by an auditor in annual auditing does not affect the everyday routine of the organization and its people; the auditor has full control over financial statements and records. Among other advantages, annual audit is cost-effective.
> **Disadvantages** – There might be instances where the unavailability of auditor may cause unnecessary delay in audit work; due to complete audit in one sitting, chances of undetected errors and frauds are high. This is not recommended for big business houses, and the delay in the annual general meeting is sometimes due to a delay in the audit which turns out to be a major disadvantage of the annual audit.

2.4.12 Partial audit

Partial audit is done only for a specific purpose, for example, to check the receipt side or the payment side of the cash book, to check cash sale, and to check purchases or expenses only. The reason for calling out for a partial audit largely depends on the management of the organization.

2.4.13 Internal audit

Internal audit may be done by an independent person or by the employees of the company; an internal auditor may or may not be a qualified person for audit. Internal audit is continuous in nature. As per section 144 of the Companies Act, an internal auditor cannot render his services as a statutory auditor for the same company.

As per the new section 138 of the Companies Act, internal audit has been made compulsory for certain categories of companies:

- Certain classes of companies may be prescribed or shall require to appoint an internal auditor, who shall be either a chartered accountant or cost accountant or such other professional as may be decided by the board to conduct an internal audit of the functions and activities of the company.
- The central government, may, by rules, prescribe the manner and intervals in which the internal audit shall be conducted and reported to the board.

The following classes of companies are required to appoint an internal auditor –

- Listed companies.
- Unlisted companies and private companies, meeting any of the following criteria as mentioned in Table 2.2.

2.4.14 Management audit

Efforts are done to bring out an overall improvement in management efficiency through a review of all the objectives, policies, procedures, and functions of management. Only a person having good knowledge and experience of management techniques may be appointed as management auditor.

Objectives of management audit

Following are the main objectives of management audit:

- To help management in setting sound objectives.
- To ensure the fulfillment of objectives.

Table 2.2 Various criteria listed for private and unlisted companies that require to appoint an internal auditor

Criteria	Private Company	Unlisted Company
Turnover	Rs. 200 crore or more during the preceding financial year	Rs. 200 crore or more during the preceding financial year
Paid up share capital	No such criteria is applicable to private company	Rs. 50 crore or more during the preceding financial year
Outstanding deposits	No such criteria is applicable to private company	Rs. 25 crore or more at any point of time during the preceding financial year
Outstanding loans or borrowings from banks or public financial institutions	Exceeding Rs. 100 crore at any point of time during the preceding financial year	Exceeding Rs. 100 crore at any point of time during the preceding financial year.

- To give recommendations about changes in policies and procedures for better results.
- To help management in elaborating duties, rights, and liabilities of the employees.
- To help management in establishing a good and sound relationship with outsiders.

2.4.15 Post & Vouch Audit

Under this audit system, we have checking of every single original entry and their posting in ledger along with balancing and totaling. This audit system is only advisable in small business units; in big business houses, an internal auditor does this job and an auditor just checks the effectiveness of the internal control system of that organization.

2.4.16 Audit in depth

Audit in depth means detailed stepwise verification of some specific transactions; this helps an auditor to understand the complete procedure of transaction as adopted by the organization to carry out any transaction. For example, to check the purchase transaction, an auditor will check the quotations, purchase orders (P.O.), material receipt notes (M.R.N), goods/material inspection notes, bin cards, and stock ledger.

2.4.17 Interim audit

Interim audit is done between two annual audits of an organization for a part of the year. It enables the board of directors to declare an interim dividend and also to determine interim figures of sales.

2.5 AUDIT RESPONSIBILITIES

The auditor's objectives are to obtain reasonable assurance about whether the financial statements as a whole are free from material misstatement, whether due to fraud or error, and to issue an auditor's report that includes the auditor's opinion. Reasonable assurance is a high level of assurance but is not a guarantee that an audit conducted in accordance with International Standards on Auditing (UK) (ISAs (UK)) will always detect a material misstatement when it exists. Misstatements can arise from fraud or error and are considered material if, individually or in the aggregate, they could reasonably be expected to influence the economic decisions of users taken on the basis of these financial statements.

As part of an audit in accordance with ISAs (UK), the auditor exercises professional judgment and maintains professional skepticism throughout the audit. The auditor also:

- Identifies and assesses the risks of material misstatement of the entity's (or where relevant, the consolidated) financial statements, whether due to fraud or error, designs and performs audit procedures responsive to those risks, and obtains audit evidence that is sufficient and appropriate to provide a basis for the auditor's opinion. The risk of not detecting a material misstatement resulting from fraud is higher than for one resulting from error, as fraud may involve collusion, forgery, intentional omissions, misrepresentations, or the override of internal control. The auditor includes an explanation in the auditor's report of the extent to which the audit was capable of detecting irregularities, including fraud.
- Obtains an understanding of internal control relevant to the audit in order to design audit procedures that are appropriate in the circumstances but not for the purpose of expressing an opinion on the effectiveness of the entity's (or where relevant, the group's) internal control.
- Evaluates the appropriateness of accounting policies used and the reasonableness of accounting estimates and related disclosures made by the directors.
- Concludes on the appropriateness of the directors' use of the going concern basis of accounting and, based on the audit evidence obtained, whether a material uncertainty exists related to events or conditions that may cast significant doubt on the entity's (or where relevant, the group's) ability to continue as a going concern. If the auditor concludes that the use of the going concern basis of accounting is appropriate and no material uncertainties have been identified, the auditor reports these conclusions in the auditor's report. If the auditor concludes that a material uncertainty exists, the auditor is required to draw attention in the auditor's report to the related disclosures in the financial statements or, if such disclosures are inadequate, to modify the auditor's

opinion. The auditor's conclusions are based on the audit evidence obtained up to the date of the auditor's report. However, future events or conditions may cause the entity (or where relevant, the group) to cease to continue as a going concern.
- Evaluates the overall presentation, structure, and content of the financial statements, including the disclosures, and whether the financial statements represent the underlying transactions and events in a manner that achieves fair presentation (i.e., gives a true and fair view).
- Where the auditor is required to report on consolidated financial statements, he obtains sufficient appropriate audit evidence regarding the financial information of the entities or business activities within the group to express an opinion on the consolidated financial statements. The group auditor is responsible for the direction, supervision, and performance of the group audit. The group auditor remains solely responsible for the audit opinion.

The auditor communicates with those charged with governance regarding, among other matters, the planned scope and timing of the audit and significant audit findings, including any significant deficiencies in internal control that the auditor identifies during the audit. For listed entities and public interest entities, the auditor also provides those charged with governance with a statement that the auditor has complied with relevant ethical requirements regarding independence, including the FRC's Ethical Standard, and communicates with them all relationships and other matters that may reasonably be thought to bear on the auditor's independence, and where applicable, related safeguards. Where the auditor is required to report on key audit matters, from the matters communicated with those charged with governance, the auditor determines those matters that were of most significance in the audit of the financial statements of the current period and are therefore the key audit matters. The auditor describes these matters in the auditor's report unless law or regulation precludes public disclosure about the matter or when, in extremely rare circumstances, the auditor determines that a matter should not be communicated in the auditor's report because the adverse consequences of doing so would reasonably be expected to outweigh the public interest benefits of such communication.

For public interest entities, other listed entities, entities that are required, and those that choose voluntarily, to report on how they have applied the UK Corporate Governance Code, and other entities subject to the governance requirements of The Companies (Miscellaneous Reporting) Regulations 2018, the auditor is required to include in the auditor's report an explanation of how the auditor evaluated management's assessment of the entity's ability to continue as a going concern and, where relevant, key observations arising with respect to that evaluation.

2.5.1 Reporting on the financial statements

- The auditor's report is required to contain a clear expression of opinion on the financial statements taken as a whole.
- To form an opinion on the financial statements, the auditor concludes as to whether:
- The financial statements adequately refer to or describe the applicable financial reporting framework;
- The financial statements appropriately disclose the significant accounting policies selected and applied. In making this evaluation, the auditor considers the relevance of the accounting policies to the entity (or where relevant, the group) and whether they have been presented in an understandable manner;
- The accounting policies selected and applied are consistent with the applicable financial reporting framework and are appropriate;
- The accounting estimates made by the directors are reasonable;
- The information presented in the financial statements is relevant, reliable, comparable, and understandable. In making this evaluation, the auditor considers whether:
- The information that should have been included has been included, and whether such information is appropriately classified, aggregated or disaggregated, and characterized; and the overall presentation of the financial statements has been undermined by including information that is not relevant or that obscures a proper understanding of the matter disclosed;
- The financial statements provide adequate disclosures to enable the intended users to understand the effect of material transactions and events on the information conveyed in the financial statements;
- The terminology used in the financial statements, including the title of each financial statement, is appropriate.
- When the financial statements are prepared in accordance with a fair presentation framework, the auditor also evaluates whether the financial statements achieve fair presentation (i.e., gives true and fair view) including consideration of:
- The overall presentation, structure, and content of the financial statements; and
- Whether the financial statements represent the underlying transactions and events in a manner that achieves fair presentation (or gives a true and fair view).

2.5.2 Unmodified opinions

An unmodified opinion is expressed when the auditor is able to conclude that the financial statements give a true and fair view 1 and comply in all material respects with the applicable financial reporting framework.

2.5.3 Modified opinions

The auditor modifies the opinion when either:

- The auditor concludes that, based on the audit evidence obtained, the financial statements as a whole are not free from material misstatement; or
- The auditor is unable to obtain sufficient appropriate audit evidence to conclude that the financial statements as a whole are free from material misstatement.

The auditor expresses a qualified opinion when either:

- Misstatements, individually or in the aggregate, are material but not pervasive to the financial statements; or
- The possible effects on the financial statements of undetected misstatements, arising from an inability to obtain sufficient appropriate audit evidence, could be material but not pervasive.
- The auditor expresses an adverse opinion when the auditor, having obtained sufficient appropriate audit evidence, concludes that misstatements, individually or in the aggregate, are both material and pervasive to the financial statements.

The auditor disclaims an opinion when either:

- The possible effects of undetected misstatements, arising from an inability to obtain sufficient appropriate audit evidence, could be both material and pervasive to the financial statements; or
- In extremely rare circumstances involving multiple uncertainties, the auditor concludes that notwithstanding having obtained sufficient appropriate audit evidence regarding each of the individual uncertainties, it is not possible to form an opinion on the financial statements due to the potential interaction of the uncertainties and their possible cumulative effect on the financial statements.

2.5.4 Emphasizing certain matters without modifying the opinion

In certain circumstances, an auditor's report includes an emphasis of matter paragraph to draw attention to a matter presented or disclosed in the financial statements that, in the auditor's judgment, is of such importance that it is fundamental to users' understanding of the financial statements. An emphasis of matter paragraph does not modify the auditor's opinion.

2.5.5 Communicating "other matters"

If the auditor considers it necessary to communicate a matter other than those that are presented or disclosed in the financial statements that, in the auditor's judgment, is relevant to users' understanding of the audit, the auditor's responsibilities or the auditor's report, the auditor does so in a separate section in the auditor's report with the heading "Other Matter" or other appropriate headings.

2.5.6 Other information included in the annual report

The auditor is required to read all financial and non-financial information (other information) included in the annual report and to identify whether the other information is materially inconsistent with the financial statements or the auditor's knowledge obtained in the audit or otherwise appears to be materially misstated.

If the auditor identifies material inconsistencies or apparent material misstatements, the auditor determines whether there is a material misstatement in the financial statements or a material misstatement of the other information. Where the auditor concludes that there is an uncorrected material misstatement of the other information, the auditor is required to report this in the auditor's report.

For entities that report on how they have applied the UK Corporate Governance Code, the auditor reviews the directors' statement in relation to going concern, longer-term viability and that part of the Corporate Governance Statement relating to the entity's compliance with the provisions of the UK Corporate Governance Code and reports on whether they are materially consistent with the financial statements and the auditor's knowledge obtained in the audit.

2.5.7 Other legal and regulatory requirements

The auditor may be required to address other legal and regulatory requirements relating to other auditor's responsibilities in the auditor's report.

Only applicable with respect to fair presentation (or true and fair) frameworks. This conclusion is required only with respect to financial statements which have been prepared in accordance with a fair presentation (or true and fair) framework (examples are International Financial Reporting Standards as adopted by the European Union and United Kingdom Generally Accepted Accounting Practice).

2.5.8 Reporting on the financial statements

The auditor's report is required to contain a clear expression of opinion on the financial statements taken as a whole.

To form an opinion on the financial statements, the auditor concludes as to whether:

- Sufficient appropriate audit evidence has been obtained.
- Uncorrected misstatements are material, individually, or in aggregate.
- The financial statements, including the disclosures, give a true and fair view.[2]
- The financial statements are prepared, in all material respects, in accordance with the requirements of the applicable financial reporting framework, including the requirements of applicable law.

In particular, forming an opinion on and reporting on the financial statements involves evaluating whether:

- The financial statements adequately refer to or describe the applicable financial reporting framework.
- The financial statements appropriately disclose the significant accounting policies selected and applied. In making this evaluation, the auditor considers the relevance of the accounting policies to the entity (or where relevant, the group) and whether they have been presented in an understandable manner.
- The accounting policies selected and applied are consistent with the applicable financial reporting framework and are appropriate.
- The accounting estimates made by the directors are reasonable.
- The information presented in the financial statements is relevant, reliable, comparable, and understandable. In making this evaluation, the auditor considers whether:
- The information that should have been included has been included, and whether such information is appropriately classified, aggregated or disaggregated, and characterized.
- The overall presentation of the financial statements has been undermined by including information that is not relevant or that obscures a proper understanding of the matter disclosed.
- The financial statements provide adequate disclosures to enable the intended users to understand the effect of material transactions and events on the information conveyed in the financial statements.
- The terminology used in the financial statements, including the title of each financial statement, is appropriate.
- When the financial statements are prepared in accordance with a fair presentation framework, the auditor also evaluates whether the financial statements achieve fair presentation (i.e., gives true and fair view) including consideration of:
- The overall presentation, structure, and content of the financial statements.

- Whether the financial statements represent the underlying transactions and events in a manner that achieves fair presentation (or gives a true and fair view).

2.6 AUDIT TIME AND PROCESS FLOW

2.6.1 What is a process?

In order to audit a process, you must first understand what it is which is diagrammatically represented in Figure 2.1. Chemical engineering could more appropriately be called process engineering. The very first engineering class that I took was called "Process Principles" as mentioned in Table 2.3.

It was considered a difficult class and the demands of the class caused many students to explore alternate career paths. In this beginner class, the students were taught about designing processes, determining duties to be performed, establishing specifications and requirements, and integrating the various units (activities) into a coordinated plan. Additionally, we were told that "problems cannot be segregated and each treated individually without consideration of the others"[1]. So the first principle we learned was that process activities are connected or linked. Secondly, we learned that processes are responsible for the changes that take place within a system. Some call this changing a transformation.

The balancing and equilibrium of inputs and outputs is called the **Law of Conservation**. The Law of Conservation requires that the sum of the inputs equal the sum of the outputs for a given process. For example, the uncut metal plate input equals the fabricated bracket, scrap, and metal filings.

The third principle is that the Law of Conservation applies to a defined process. For processes to work, inputs and outputs must be in balance. If process elements are out of balance, the objectives would not be achieved and the process would not be effective. The output objective demands certain inputs, and if you don't have sufficient inputs, the outputs will never be achieved.

A fourth principle is that processes can be operated at a set of optimum conditions for the best utilization of resources and achievement of objectives. Economics is an important consideration in design and process operation. Every process has a set of optimum operating conditions for achieving both economic and performance objectives.

2.6.2 Process description

A process transforms inputs into outputs. This transformation or change takes place as a series of activities or steps that lead to the desired result (objective). The "process approach" way of doing things is more effective in achieving objectives than a haphazard or random approach. Establishing

Audit Planning and Preparation 53

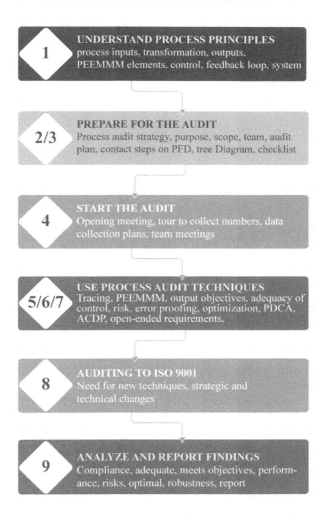

Figure 2.1 A step-by-step process to explain in detail auditing which helps an individual to clearly understand the process of auditing.

a process is good, but it could be either a good process or a bad process, similar to the thought captured in the saying "A bad plan is better than no plan".

Process inputs can be tangible or intangible. Inputs may come in the form of people, equipment, materials, parts, assemblies, components, information, money, and so on, as mentioned in Figure 2.2. The process

Table 2.3 The Process Principles

Process	Principles
1	Activities are linked as sequential steps
2	Change (transformation) takes place
3	The Law of Conservation applies to a defined process
4	Optimization results in the best utilization of resources

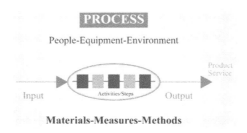

Figure 2.2 Process equilibrium where inputs, the inputs can be people, information, money, or things, and the outputs can be a product(s) or service(s).

transforms, changes, or converts the inputs into an output. Inputs are the thing(s) that will be transformed and the transformation mechanisms. An output may be a product or service. Outputs can include parts, assemblies, materials, information, energy, money, machines, devices, equipment, completed treatment, and performance of a skill. The output product is a result of the process.

There are many other inputs to a process (other than the object of the transformation) that make it possible to complete the transformation. It is convenient to categorize these inputs as process elements. The process elements can be divided into six groups that contain all the factors that make up a process.

The six groups are **people, equipment, environment, materials, measures, and methods** (PEEMMM, see Tables 2.4 and 2.5).

If the process is making a peanut butter and jelly sandwich, we will be transforming the materials peanut butter, jelly, and bread to a sandwich. Others use different terms to describe the same thing.

In a machine shop, there may be a process for making brackets that transforms a metal plate into H brackets.

In all cases, there must be sufficient inputs to achieve the output. There must be a balance or equilibrium between inputs and outputs (Law of Conservation).

For example, if you want to make 1000 H brackets, there has to be the required amount of metal, energy, and machine time. If inputs are too few, too many, or different from the process needs, the process will be suboptimal.[3] Too few inputs will result in a shortfall; too many will result in

Table 2.4 A sample case study to illustrate six factors namely, people, equipment, environment, materials, measures, and methods (PEMMM) which together form a process

Question	Answer	Explanation
Are people involved in this process?	Yes	We need a chef
Do we need equipment to complete this process?	Yes	A knife to spread the ingredients and perhaps a cutting board
Are there any environmental considerations?	Yes	The area needs to be clean. There should not be any water on the cutting board or countertop that the bread might soak up.
Are materials needed?	Yes	Grape jelly, bread, and crunchy peanut butter.
Are there measures or standards?	Yes	We need to ensure the sandwich consumer is pleased with outcome.
Do we need a method?	Yes	We will need a recipe that will tell the steps, order, and amounts.

Table 2.5 A case study on "H Bracket Stamping Process" for better understanding of PEMMM and validating the existence of PEMMM

EXAMPLE: H Bracket Stamping Process
People – Equipment – Environment – Materials – Measures – Methods

Question	Answer	Explanation
Are people involved in this process?	Yes	Machine operator, material helper
Do we need equipment to complete this process?	Yes	250-ton stamping machine, tote bin, scrap cart, tags, safety equipment
Are there any environmental considerations?	Yes	Housekeeping: (look for excessive dust, no oil on floor)
Are materials needed?	Yes	Machine oil, cut and sized metal plate
Are there measures or standards?	Yes	Check dimensions with calipers per traveler
Do we need a method?	Yes	Traveler (work order), work instruction 075-H-02, safety manual

excessive waste; and anything other than the required inputs will result in an ineffective process and/or chaos.

2.6.3 Control of processes

In most cases, it is desirable to control processes to avoid negative consequences. The amount or level of control varies depending on the risk and acceptability of undesirable outcomes. The severity or level of control for nuclear plant processes will be different from that of any organization that publishes a magazine or newspaper. The essential requirement for any control system is that there is a feedback information loop from the process output. The feedback information is used to adjust the process or make decisions about the output (before the next process or before the customer receives it). Feedback information may be in the form of temperature, pressure, dimension, weight, volume, count, color, condition, or portion. The function of the control feedback loop is to achieve output targets and objectives.

For management to control a process or activity, it must establish a predetermined method. Without it, there is no basis to adjust or improve the process. Predetermined methods can include plans, procedures, work instructions, checklists, outlines, diagrams, flow charts, step-by-step software program code, process maps, and so on.

Feedback information should relate to the process performance criteria and/or objectives. Feedback can be in the form of a quality characteristic such as activity level or dimensions, and feedback can be a performance measure such as yield, cost, waste, delays, utilization, error rate, field failures, satisfaction, and so on. Sometimes it is easier to monitor a process parameter such as temperature or pressure that is a function or indicator of the performance of the process. For example, we can monitor the temperature of a process because we know that if it goes above 107°C, the product will darken and fail the color test. Without a feedback loop, objectives cannot be assured.

For example (Figure 2.3), there may be a process that transforms sheet metal into an H bracket with specified holes for fasteners. One of the most critical parameters may be the diameter of the holes. For this process, the feedback loop is the diameter of the holes. This information can be used to

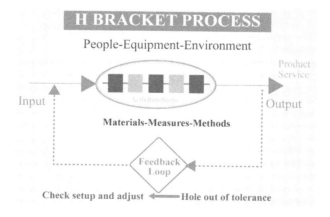

Figure 2.3 H bracket process control illustrates the transformation of sheet metal into an H bracket with specified holes for fasteners and the feedback loop is the diameter of the holes.

reject parts or make adjustments to the fabrication process (machine, drill bits, and so on).

In some cases, you may want to use statistical techniques to better understand what changes need to be made to the process to ensure that output objectives are achieved. For example, Statistical Process Control (SPC) charts may be used to monitor a process variable to ensure the process is capable and out-of-control points are acted upon.

Processes must be organized with other processes to achieve objectives such as manufacturing and distributing parts for a profit. Processes are sequenced and linked to achieve certain business/organization objectives. An organization can be considered a collection of processes, all of which are working together to transform inputs into outputs.

The combination of product/service processes with management processes creates a system. Processes can be simple or complex, similar or dissimilar. The system brings the processes together for a common purpose that may relate to business, services, quality, environment, or safety.

2.6.4 Advanced process and system modeling

There are other palatable modeling techniques. The class has information about what is called the IDEF3 Process Flow and Objective State Description Capture Method Overview. IDEF stands for Integrated DEFinition. The IDEF methodology may be used to model a wide variety of automated and non-automated "systems" or subject areas, including any combination of hardware, software, machines, processes, or people. The IDEF modeling techniques can be used to model very complex systems and processes. This model uses the term mechanisms to describe the people and machines that do the transformation work.

2.7 ISMS AUDIT CHECKLIST

2.7.1 Why ISO 27001 Checklist is required? What is the importance of ISO 27001 Checklists?

- If a business is worth doing, then it is worth doing it in a secure manner. Hence, there cannot be any compromise. Without a comprehensive professionally drawn information security checklist by your side, there is the likelihood that compromise may take place. This compromise is extremely costly for organizations and professionals.
- Information security management audit is though very logical requires a systematic detailed investigative approach. For a newbie entity (organization and professional), there are proverbial many a slips between cup and lips in the realm of information security management thorough understanding, let alone ISO 27001 Audit.

- Even with several years of experience by an entity's (organization and professional) side, information security assessments (read investigations) go astray due to several reasons including engineered distractions, bias, time constraints, (un)comfortable niches, auditee-guided audit (investigation), and lack of optimum exposure and experience.
- With each vulnerability/risk at the organization level, site level, department level, process and sub-process level, device and component level, tools/application level, people level, technology platform level, and delivered products/services level, it is humanly possible to miss out a large number of unidentified vulnerabilities/risk due to various reasons including ignorance, rush, vested disinterest, insider threat, connivance between the various working groups, tendency to promote tools for shear commercial interests rather than a holistic security solution, and so on, and the list is very long. Comprehensive and detailed ISO 27001 Checklist Questions enable "carpet bombing" of all ISMS requirements to detect what "exactly" is the compliance and non-compliance status.
- What is the biggest risk for an organization? The biggest vulnerability is the "gang of unidentified risks", lurking in the dark, ready to pounce when the victim organization least expects it. The risks in this gang work sympathetically and in synergy to inflict maximum damage, including corporate mortality, huge penalties by the customers/clients and regulatory bodies, flight away of business, loss of reputation and brand value, loss of jobs, and bankruptcy. This becomes very much possible without a professionally drawn comprehensive and robust ISO 27001 Checklist by your side.
- Of course, an Information Security Audit becomes a robust, immensely focused, efficient, time saver exercise with sharp ISO 27001 Checklist Questions.

2.7.2 Who all can use ISO 27001 Audit Checklist?

These detailed Information Security Compliance Audit Questions checklists are useful for:

1. Organization planning for ISO 27001 Certification.
2. Compliance audits.
3. Gap assessments prior to mergers and acquisitions, ISO 27001 Certification audit, and vendor selection due diligence.
4. Enhancing longevity of the business by helping to conduct business in the most secure manner.
5. Organizations keen on robust, resilient, and value-added Information Security Management System.
6. Organizations keen to protect themselves against entire ISMS framework issues from requirements of ISO 27001.

Audit Planning and Preparation 59

7. Organizations that want to survive client audits.
8. Information security professionals.
9. Internal auditors of Information Security Management System.
10. External auditors of Information Security Management System.
11. Auditors of the client organizations tasked to assess the ISMS capability of their service providers, vendors, and contractors.
12. Students of Information Security Management System.
13. ISO 27001 lead auditor training participants.
14. ISO 27001 lead implementer participants.
15. Professionals doing career switchover to information security.
16. Owners of business.
17. Chief Technology Officer (CTO), CIO, Chief Information Security Officer, head of the department (HODs), ISO 27001 SPOCs from departments, IT teams, central security team.

2.7.3 How many ISO 27001 Checklists are available?

1. There are two catalogs of ISO Checklists, namely, "Clause-Wise Checklist for ISO 27001" and "Department-Wise Checklist for ISO 27001".
2. Clause-wise audit checklist spans all the clauses of the Information Security Management System framework, i.e., Clauses 4 to 10.2 are auditable clauses. Clauses 1 to 3 are non-auditable clauses and therefore not covered in the clause-wise checklist.
3. Department-wise checklists cover all critical verticals such as the IT department, Software Design and Development department, HR and Training department, and Admin department.

2.7.4 How to find out which ISO 27001 Checklists are suitable for me?

2.7.4.1 For an organization aiming for ISO 27001 Certification

1. Whether aiming for ISO 27001 Certification for the first time or maintaining ISO 27001 Certificate, vide periodical surveillance audits of ISMS; both clause-wise checklist and department-wise checklist are suggested and compliance audits can be performed as per the checklists.
2. The same holds good for performing supplier (second-party) audits, internal (first-party) audits, and external (third-party, certification) audits.
3. Information security consultants would require both clause-wise checklists and department-wise checklists.

2.7.4.2 For a head of the department?

HODs should focus on the checklist of their respective departments only, for example, the head of the Human Resource department should only focus on "ISO 27001 HR Audit Checklist|272 Compliance Questions".

2.7.4.3 For a CISO (Chief Information Security Officer)

CISOs should focus on the ISMS framework checklist, i.e., "ISO 27001 Audit Checklist – Clauses 4 to 10.2 – 1336 Questions". CISO may ensure the organization obtains a critical department's checklist centrally.

2.7.4.4 For a CTO (Chief Technology Officer) and CIO

CTO and CIO should focus on all IT department checklists and "Information Security Risk Management Audit Checklist|Clauses 6.1.1, 6.1.2, 6.1.3, 8.2, 8.3 – 251 Questions".

2.7.4.5 For IT department professionals

Look for your domain-specific checklists, for example, network security, IT help desk security, and Web security, should focus on "Information Security Risk Management Audit Checklist|Clauses 6.1.1, 6.1.2, 6.1.3, 8.2, 8.3 – 251 Questions".

2.7.4.6 For preparing for a job interview

Look for your weak areas and strengthen them with help of checklist questionnaires. The thumb rule is to make your niches strong with help of a niche/vertical specific checklist. Key point is to walk the talk with the Information Security Management System in your area of operation to land yourself your dream assignment. Everything being equal, functional knowledge + information security exposure will surge you ahead of your competitors. Look for your domain-specific checklists, for example, network security, IT help desk security, Web security, etc.

2.7.5 Important information on ISO 27001 Checklist file

- File format: Excel compatible for both Mac and Windows.
- Contains: as described in the description mentioned above.
- Language: English.
- File delivery method: immediate and automatic. Through the secure link in the email provided at the time of check-out.

- Link validity: 72 hours from the time of receiving the link through email.
- Invoice: invoice is generated on your device immediately after successful payment.

2.7.6 Who has prepared and who has validated ISO 27001 Checklists?

These ISO 27001 Checklists are prepared by an expert panel of IRCA Principal Auditors and Lead Instructors of Information Security Management System having aggregated panel team experience of over 300 years, under the aegis of ISO Training Institute. The checklist is validated by the head of the expert committee and approved by ISO Training Institute.

2.7.7 What is the basis of the ISO 27001 Checklist?

The Information Security Audit checklist on Requirements of ISO 27001 follows the cardinal of:

- Risk-based thinking (RBT).
- Process approach.
- PDCA (Plan Do Check Act) methodology.

The expert panel of information security auditors and instructors has conducted thousands of information security audits and training on ISO 27001. Besides, there is a continuous calibration of the lead auditors w.r.t requirements, interpretation, and audit experiences.

2.7.8 How to use ISO 27001 Checklist?

- Securely save the original checklist file and use the copy of the file as your working document during the preparation/conduct of the Information Security Audit.
- The organization's InfoSec processes are at varying levels of ISMS maturity; therefore, use checklist quantum apportioned to the current status of threats emerging from risk exposure.

BIBLIOGRAPHY

1. Case.edu., 2022. *Audit plan & process | Audit services.* Case Western Reserve University. [online] Available at: <https://case.edu/auditservices/audit-plan-process> [Accessed 26 February 2022].

2. 27001Academy, 2022. *ISO 27001 & ISO 22301 academy: Online tools, advice & certification.* [online] Available at: <https://advisera.com/27001academy/> [Accessed 26 February 2022].
3. 2022. [online] Available at: <https://www.isocertificationtrainingcourse.org/iso-27001-checklist> [Accessed 26 February 2022].
4. Calder, A., 2017. *Nine steps to success: An ISO 27001 implementation overview.* IT Governance Ltd.
5. Kenyon, B., 2019. *ISO 27001 controls: A guide to implementing and auditing.* IT Governance Ltd.
6. Perera, D., 2008. *ISO/IEC 27001, information security management system.*

Chapter 3
Audit Techniques and Collecting Evidence

3.1 AUDITOR QUALITY AND SELECTION

Companies are going to change their auditors more frequently than in the past. A good selection process is essential to enhance audit quality and ensure auditor's independence. Therefore, here, a four-step approach is provided for a high-quality selection process. We previously identified these steps for auditor selection: Toward best practices which was based on a pan-European stakeholders' survey and the work of an expert task force. This guidance will benefit (1) companies in managing the auditor selection process and (2) shareholders in assessing a company's current selection process.

It can be of use to select auditors irrespective of the regulatory environment, country, market segment, or size of the company – e.g., public-interest entities (PIEs) and small- and medium-sized enterprises (SMEs). Due to the diverse legal frameworks and governance structures across Europe, the reader may also need to refer to local company law and corporate governance requirements.

How to prepare for an auditor selection process

Preparation is the key to success when it comes to choosing an auditor who fulfills the company's needs. There is value in investing in the project and mitigating the risks of selecting an auditor who is not fit for the purpose.

The following points could be taken into account in preparing the selection process:

- Plan for the time required by the process.
- Develop a step-by-step procedure. The process should be robust and proportionate using project management techniques communicate the primary objectives and the company's expectations to those invited to tender and take an open-book approach so that the same information is available to all auditors who participate in the process.

DOI: 10.1201/9781003322191-3

- For those invited to tender, consider the following questions: Should the incumbent auditor be invited to tender? How many auditors are to be invited? Because of time constraints, we recommend that no more than six firms be invited to tender for the short list and two would be an adequate number.
- Plan the transition of auditors in advance.

Four steps to select an auditor

We recommend a step-by-step procedure to be put in place to prepare for selecting an auditor. We propose the following four steps in the selection process:

- Information-gathering.
- Pre-screening of possible auditors.
- Development of criteria to select the auditor.
- Ranking of the chosen criteria.

Step 1: Information-gathering

Governance aspects of the appointment of the auditor from a legal perspective, the auditor is formally appointed/elected by the shareholders of a company or by those charged with the entity's governance. The formal appointment is usually based on the recommendation of the board or the audit committee. A different governance approach might be useful depending on the entity. For public-interest entities (PIEs) and other large entities, involving independent members of the board or the audit committee, if applicable, could be relevant for PIEs. For smaller unlisted companies, it could be beneficial if directors and shareholders become more involved in the process. The relationship between an auditor and the audited company is much more than a pure "buyer–seller" one. The audit has historically been performed to serve shareholders. Now it is more and more about serving investors and, ultimately, the public at large. Audit quality, as a shared commitment between all players in the financial reporting supply chain, should be at the center of the discussion. Therefore, the decision to "purchase" the audit service should be based on an overall assessment of all relevant criteria and not only the economic criteria. Certain corporate governance frameworks even prohibit the involvement of executives of the company in the auditor selection process.

Step 2: Pre-screening of possible auditors

Determining the needs of the company. The company's needs have to be properly defined before starting the auditor selection process. These needs can be linked to regulatory requirements or to a need to assess the current audit approach and the service provided. The company may also reflect on whether it is the right moment for the company to initiate this process. Independence of the auditor A letter of representation has to be obtained. It is a letter from the potential auditor that includes information on whether the auditor is independent of the company and how the auditor monitors and maintains his/her independence. The company should review the audit-related services and non-audit services of the candidate auditors over the past years and analyze compliance with any other independence requirements.

Step 3: Development of criteria to select the auditor

Depending on the needs and objectives defined in the previous step, the evaluation criteria may include the following:

- Approach to business and operations in general business model and governance of the audit firm to internal processes to ensure independence and other relevant rules are correctly applied.
- Audit approach description of the methodology to be used by the auditor areas that will receive primary emphasis and the audit approach in such areas comprehensive work plan to ensure an adequate coverage business understanding industry-specific experience, if applicable use of its tools, use of associated or affiliated member firm personnel and third-party experts, if necessary.
- Communication strategy additional internal status report, i.e., to outline weaknesses in internal controls means put in place to ensure the timeliness of the information policy regarding the availability of partners and managers for miscellaneous inquiries and short meetings throughout the year.
- Reputation: Auditors need to demonstrate a good and ethical reputation. Requesting references may help in this evaluation.
- Evidence of audit quality: Need to understand the auditor's system of internal quality assurance based on the auditor's presentation and available documentation.
- Assessment of the individual auditor/audit engagement partner: The individual auditor/audit engagement partner(s) put forward to be in charge of the audit should be met in person. It is a good opportunity to assess whether they meet the professional expectations about the firm as anticipated in the tender documents.
- People management qualification – audit qualifications of the team members – involvement of experts on specific subject matters, such as tax issues and actuarial services (such involvement will enhance audit quality depending on sector specificities) training – continuing professional development (CPD) experience – expertise and knowledge of audit engagement partner – appropriate level of seniority of team members and effective contribution at the relevant level – informed audit team with international outreach, where necessary – relevant industry experience and expertise of the audit firm and/or the audit

team availability – availability of the engagement partner – staff continuity, i.e., staff turnover records from previous years.
- Geographical coverage: Important for audits of multinational entities.
- Insurance coverage: The auditor should comply with what is required by local legislation or justified by the needs of the business, the sector, and the circumstances.
- Pricing: The audit contract should be awarded on the basis of the most economically advantageous offer, not the lowest price. This is to ensure that the quality of the audit is not compromised to identify the most economically advantageous offer, audit fees should be assessed against the availability of the key members of the team and the resources of the audit firm as whole personnel resources, their expertise and qualifications allocation of personnel, i.e. hours to be spent allocated to each type and level of qualified resource risk approach and the audit methodology – these can have a significant effect on pricing for both sides (e.g. gain in efficiency, use of experts, site visits).
- Relationship management and interpersonal skills: The auditor applying should be able to demonstrate capacity in building objective, independent, and transparent working relationships with the management team and task forces, as well as with those charged with governance. A good relationship and positive attitude are not indicative of a lack of independence or conflict of interest on the part of the auditor. The right balance has to be struck between professional skepticism and cooperation.
- Capacity for innovation: The auditor should be able to demonstrate his/her ability to improve the audit processes, for instance through the use of technology step.

Step 4: Ranking of the chosen criteria

Different criteria have to be taken into account when performing the evaluation. Some may be easy to measure (e.g., factual criteria), while others may be more subjective (e.g., soft-skill criteria). Not all evaluation criteria are strategic for all entities. Their importance varies depending on the company; the criteria relevant to the audited entity need to be properly weighted.

3.2 AUDIT SCRIPT

The standard audit scripts confirm that modifications controlled by finish scripts were made to the system, and they report any discrepancies that occurred since the hardening run. Audit scripts use the same name as their correlating finish script, except they have a different suffix. Audit scripts use the aud suffix instead of fin.

Customizing audit scripts

This section provides instructions and recommendations for customizing existing audit scripts or creating new audit scripts. In addition, guidelines are provided for using audit script functions.

Customize standard audit scripts

Just as with Solaris Security Toolkit drivers and finish scripts, you can customize audit scripts. Make as few changes as necessary to the code whenever possible and document those changes. Use environment variables to customize an audit script. The behavior of most scripts can be significantly altered through environment variables, thereby eliminating the need to modify the script's code directly. If this is not possible, you may find it necessary to modify the function by developing a customized one for use by the user. Run script. For a list of all environment variables and guidelines for defining them.

To customize an audit script

Use the following steps to customize a standard audit script for your system and environment. Use these instructions so that newer versions of the original files do not overwrite your customized versions. Note that these files are not removed if you use the program command to remove the Solaris Security Toolkit software.

- Copy the audit script and related files that you want to customize.
- Rename the copies with names that identify the files as custom scripts and files.
- Modify your custom script and files accordingly.

The finish.init file provides all audit script configuration variables. You can override the variable's default value specified in the finish.init file by adding the variable and its correct value to the user.init file. This file is heavily commented on to explain each variable, its impact, and its use in audit scripts. If you want the change to be localized rather than to apply to all drivers, modify the driver. When you customize audit scripts, it is critical to the accuracy of the audit functionality that both finish and audit scripts are able to access your customization. This goal is most easily and effectively achieved by modifying environment variables in the user.init script instead of modifying other init files or modifying scripts directly.

Code Example 3-1 shows how the install-openssh.audit script validates a correct software installation by checking whether the software package is installed, configured, and set up to run whenever the system reboots. In this example, these checks ensure that the software package is installed, configured, and set up to run whenever the system reboots.

Using standard audit scripts

Audit scripts provide an automated way within the Solaris Security Toolkit software to validate a security posture by comparing it to a predefined security profile. Use audit scripts to validate that security modifications were made correctly and to obtain reports on any discrepancies between a system's security posture and your security profile. This section describes the standard audit scripts, which are in the Audit directory. Following is a code example for the audit script:

Code Example 3.1 Sample install-openssh.aud Script

```
#
#!/bin/sh
#ident "@(#)install-openssh.aud
# Service definition section. service="OpSH"
  servfil="install-openssh.aud" servhdr_txt="
#Rationale for Verification Check:
#This script will attempt to determine if the OpenSSH
  software is
#installed, configured and running on the system. Note
  that this
#script expects the OpenSSH software to be installed in
  package
#form in accordance with the install-openssh.fin Finish
  script.
#Determination of Compliance:
#It indicates a failure if the OpenSSH package is not
  installed,
#configured, or running on the system."
servpkg=" OBSDssh "
servsrc=" ${JASS_ROOT_DIR}/etc/rc3.d/S25openssh.server "
servcfg=" ${JASS_ROOT_DIR}/etc/sshd_config"
"servcmd=" /opt/OBSDssh/sbin/sshd "
# ***********************************************************
# Check processing section.
# ***********************************************************
start_audit "${servfil}" "${service}" "${servhdr_txt}"
  logMessage"${JASS_MSG_SOFTWARE_INSTALLED}"

if check_packageExists "${servpkg}" 1 LOG; then
pkgName="`pkgparam -R ${JASS_ROOT_DIR} ${servpkg} NAME`"
pkgVersion="`pkgparam -R ${JASS_ROOT_DIR} ${servpkg}
   VERSION`"
pkgBaseDir="`pkgparam -R ${JASS_ROOT_DIR} ${servpkg}
   BASEDIR`"
```

```
    pkgContact="`pkgparam -R ${JASS_ROOT_DIR} ${servpkg}
    EMAIL`"

    logNotice "Package has description '${pkgName}'"
    logNotice "Package has version '${pkgVersion}'"
    logNotice "Package has base directory '${pkgBaseDir}'"
    logNotice "Package has contact '${pkgContact}'"

    logMessage "\n${JASS_MSG_SOFTWARE_CONFIGURED}"
    check_startScriptExists "${servsrc}" 1 LOG
    check_serviceConfigExists "${servcfg}" 1 LOG


    logMessage "\n${JASS_MSG_SOFTWARE_RUNNING}"
    check_processExists "${servcmd}" 1 LOG
    finish_audit
```

Only the functionality performed by the audit scripts is described. Each of the scripts in the Audit directory is organized into the following categories, which mirror those of the finish scripts in the Finish directory:

- disable
- enable
- install
- minimize
- print
- remove
- set
- update

Disable Audit Scripts: To disable audit, we have various options. For disabling audit, we are describing following scripts:

- disable-ab2.aud
- disable-apache.aud
- disable-apache2.aud
- disable-appserv.aud
- disable-asppp.aud
- disable-autoinst.aud
- disable-automount.aud
- disable-dhcpd.aud
- disable-directory.aud
- disable-dmi.aud
- disable-dtlogin.aud
- disable-face-log.aud

- disable-IIim.aud
- disable-ipv6.aud
- disable-kdc.aud
- disable-keyboard-abort.aud
- disable-keyserv-uid-nobody.aud
- disable-ldap-client.aud
- disable-lp.aud
- disable-mipagent.aud
- disable-named.aud
- disable-nfs-client.aud
- disable-nfs-server.aud
- disable-nscd-caching.aud
- disable-picld.aud
- disable-power-mgmt.aud
- disable-ppp.aud
- disable-preserve.aud
- disable-remote-root-login.aud
- disable-rhosts.aud
- disable-routing.aud
- disable-rpc.aud
- disable-samba.aud
- disable-sendmail.aud
- disable-slp.aud
- disable-sma.aud
- disable-snmp.aud
- disable-spc.aud
- disable-ssh-root-login.aud
- disable-syslogd-listen.aud
- disable-system-accounts.aud
- disable-uucp.aud
- disable-vold.aud
- disable-wbem.aud
- disable-xfs.aud
- disable-xserver.listen.aud

Create new audit scripts

You can create new audit scripts and integrate them into your deployment of the Solaris Security Toolkit software. The same conventions for developing new finish scripts apply to developing new audit scripts. Followings are a few audit scripts:

- disable-apache.aud: This script determines if the Apache Web Server is installed, configured, or running on the system. It indicates a failure if the software is installed, configured to run, or running on the system.

- disable-apache2.aud: This script determines if the Apache 2 service is installed, configured, or running on the system. It indicates a failure if the software is installed, configured to run, or running on the system.
- disable-appserv.aud: This script determines if the Sun Java Application Server is installed, configured, or running on the system. The script indicates a failure if the software is installed or configured to run.
- disable-asppp.aud This script determines if the ASPPP service is installed, configured, or running on the system. It indicates a failure if the software is installed, configured to run, or running on the system.
- disable-autoinst.aud: This script determines if automated installation functionality is installed or enabled on the system. It indicates a failure if the software is installed or configured to run.
- disable-automount.aud: This script determines if the automount service is installed, configured, or running on the system. It indicates a failure if the software is installed, configured to run, or running on the system.
- disable-dhcpd.aud: This script determines if the DHCP service is installed, configured, or running on the system. It indicates a failure if the software is installed, configured to run, or running on the system.
- disable-directory.aud: This script determines if the Sun Java System Directory service is installed, configured, or running on the system. It indicates a failure if the software is installed, configured to run, or running on the system.
- disable-dmi.aud: This script determines if the DMI service is installed, configured, or running on the system. It indicates a failure if the software is installed, configured to run, or running on the system.
- disable-dtlogin.aud: This script determines if the CDE login server, or dtlogin, is installed, configured, or running on the system. It indicates a failure if the software is installed, configured to run, or running on the system.
- disable-face-log.aud: This script verifies that the /usr/oasys/tmp/TERRLOG file is present and has no write permissions for Group and Other. The script indicates a failure if the file has global write permissions by Group or Other.
- disable-IIim.aud: This script determines if the IIim service is installed, configured, or running on the system. The script indicates a failure if the software is installed, configured to run, or actually running on the system.
- disable-ipv6.aud: This script checks for the absence of the IPv6 host name files, /etc/hostname6.*, that cause IPv6 interfaces to be plumbed. This script checks if the in.ndpd service is started. It indicates a failure if any IPv6 interfaces are configured, plumbed, or if the service is running.
- disable-kdc.aud: This script determines if the KDC service is installed, configured, or running on the system. It indicates a failure if the software is installed, configured to run, or running on the system.

- disable-keyboard-abort.aud: This script determines if the system is configured to ignore keyboard abort sequences. Typically, when a keyboard abort sequence is initiated, the operating system is suspended and the console enters the OpenBoot PROM monitor or debugger. This script determines if the system can be suspended in this way.
- disable-keyserv-uid-nobody.aud: This script determines if the keyserv service is not configured to prevent the use of default keys for the user nobody. This script indicates a failure if the keyserv process is not running with the -d flag and the ENABLE_NOBODY_KEYS parameter is not set to NO.
- disable-ldap-client.aud: This script determines if the LDAP client service is installed, configured, or running on the system. It indicates a failure if the software is installed, configured to run, or running on the system.
- disable-lp.aud: This script determines if the line printer (lp) service is installed, configured, or running on the system. It indicates a failure if the software is installed, configured to run, or running on the system. This script also indicates a failure if the lp user is permitted to use the cron facility or has a crontab file installed.
- disable-mipagent.aud: This script determines if the Mobile IP service is installed, configured, or running on the system. It indicates a failure if the software is installed, configured to run, or running on the system.
- disable-named.aud: This script determines if the DNS server is installed, configured, or running on the system. This script indicates a failure if the software is installed, configured to run (through a configuration file), or actually running on the system.
- disable-nfs-client.aud: This script determines if the NFS client service is configured or running on the system. It indicates a failure if the software is configured to run or is running on the system.
- disable-nfs-server.aud: This script determines if the NFS service is configured or running on the system. It indicates a failure if the software is configured to run or is running on the system.
- disable-nscd-caching.aud: This script determines if any of the passwd, group, host, or ipnodes services have a positive time-to-live or negative time-to-live value that is not set to 0. The script indicates a failure if the value is not 0.
- disable-picld.aud: This script determines if the PICL service is installed, configured, or running on the system. It indicates a failure if the software is installed, configured to run, or running on the system.
- disable-power-mgmt.aud: This script determines if the power management service is installed, configured, or running on the system. It indicates a failure if the software is installed, configured to run, or running on the system.

Audit Techniques and Collecting Evidence   73

- disable-ppp.aud: This script determines if the PPP service is installed, configured, or running on the system. It indicates a failure if the software is installed, configured to run, or running on the system.
- disable-preserve.aud: This script determines if the preserve functionality is enabled. If enabled, a failure is indicated.
- disable-remote-root-login.aud: This script determines and indicates a failure, if a root user is permitted to directly log in to or execute commands on a system remotely through programs using /bin/login, such as telnet.
- disable-rhosts.aud: This script determines if the rhosts and hosts.equiv functionality is enabled through PAM configuration.
- disable-routing.aud: This script determines if routing, or packet forwarding, of network packets from one network to another is disabled.
- disable-rpc.aud: This script determines if the RPC service is installed, configured, or running on the system. It indicates a failure if the software is installed, configured to run, or running on the system. In addition, this script indicates a failure for each service registered with the rpcbind port mapper.
- disable-samba.aud: This script determines if the Samba service is installed, configured, or running on the system. It indicates a failure if the software is installed, configured to run, or running on the system. Only Samba services included in the Solaris OS distribution are verified as being disabled. This script does not impact other Samba distributions installed on the system.
- disable-sendmail.aud: By default, the sendmail service is configured to both forward local mail and receive incoming mail from remote sources. If a system is not intended to be a mail server, then the sendmail service can be configured not to accept incoming messages. This script checks that the sendmail service is configured not to accept incoming messages.
- disable-slp.aud: This script determines if the SLP service is installed, configured, or running on the system. It indicates a failure if the software is installed, configured to run, or running on the system.
- disable-sma.aud: This script determines if the SMA service is installed, configured, or running on the system. This script indicates a failure if the software is called, configured to run, or actually running on the system.
- disable-snmp.aud: This script determines if the SNMP service is installed, configured, or running on the system. It indicates a failure if the software is installed, configured to run, or running on the system. This script does not verify whether third-party SNMP agents are functioning on the system.
- disable-spc.aud: This script determines if the Statistical Process Control (SPC) service is installed, configured, or running on the

system. It indicates a failure if the software is installed, configured to run, or running on the system.
- disable-ssh-root-login.aud: This script indicates a failure if the Solaris Secure Shell service distributed in the Solaris OS versions 9 and 10 does not restrict access to the root account. disable-syslogd-listen.aud The script sets options to disallow the remote logging functionality of the syslogd process. This script determines if the SYSLOG service is configured to accept remote log connections.
- disable-system-accounts.aud: This script indicates a failure if the shell program listed in the JASS_SHELL_DISABLE variable does not exist on the system.
- disable-uucp.aud: This script determines if the UUCP service is installed, configured, or running on the system. It indicates a failure if the software is installed, configured to run, or running on the system.
- disable-vold.aud: This script determines if the VOLD service is installed, configured, or running on the system. It indicates a failure if the software is installed, configured to run, or is running on the system.
- disable-wbem.aud: This script determines if the WBEM service is installed, configured, or running on the system. It indicates a failure if the software is installed, configured to run, or running on the system.
- disable-xfs.aud: This script determines if the xfs service is installed, enabled, or running on the system. This script indicates a failure if the software is enabled to run or actually running on the system.
- disable-xserver.listen.aud: This script indicates a failure if the X11 server is configured to accept client connections using the TCP transport. In addition, it indicates a failure if the X11 server is running in a configuration that permits the use of the TCP transport.

Enable audit scripts

The following enable audit scripts are described in this section:
- enable-account-lockout.aud
- enable-bart.aud
- enable-bsm.aud
- enable-coreadm.aud
- enable-ftp-syslog.aud
- enable-ftpaccess.aud
- enable-inetd-syslog.aud
- enable-ipfilter.aud
- enable-password-history.aud
- enable-priv-nfs-ports.aud
- enable-process-accounting.aud
- enable-rfc1948.aud
- enable-stack-protection.aud
- enable-tcpwrappers.aud

Audit Techniques and Collecting Evidence   75

Here is a detailed description of performing the task of enable audit:

- enable-account-lockout.aud: This script verifies that the value of LOCK_AFTER_RETRIES is defined correctly in the /etc/security/policy.conf file. In addition, this script checks to ensure that no users have a different value than LOCK_AFTER_RETRIES specified in /etc/ user_attr.
- enable-bart.aud: This script verifies that BART has been run and compares BART rules and manifests files. The script determines if a BART rules file is present, and if so, determines if its configuration is consistent with the driver being run and its BART rules file. If the BART rules file configuration is not consistent with the driver being run and its BART rules file, the script copies a rules file from $JASS_FILES/var/opt/SUNWjass/bart/rules. The script also reports any differences between the new and most recent manifest files, generates audit messages containing the names of the BART manifests used, and suggests that the user check against earlier manifest files or the FingerPrint Database for any issues found.
- enable-bsm.aud: This script determines if the SunSHIELD Solaris Basic Security Module (Solaris BSM) auditing functionality is enabled and running on the system, if the service is loaded in the /etc/ system file, and if the audit_warn alias is defined in /etc/mail/aliases. If one or more of these checks fail, then the script indicates a failure.
- enable-coreadm.aud: This script verifies that the system stores generated core files under the directory specified by JASS_CORE_DIR. It indicates a failure if the coreadm functionality present in the Solaris OS versions 7 through 10 is not configured. An error condition also is generated if core files are not tagged with the specification denoted by JASS_CORE_PATTERN.
- enable-ftp-syslog.aud: This script determines if the FTP service is not configured to log session and connection information. A failure is indicated if the FTP service logging is not enabled.
- enable-ftpaccess.aud: This script determines if the FTP service is configured to use the /etc/ftpd/ftpaccess file. A failure is indicated if FTP is not configured properly.
- enable-inetd-syslog.aud: This script determines if the Internet services daemon (inetd) service is configured to log session and connection information.
- enable-ipfilter.aud This script reviews the ipfilter configuration of all available network interfaces and verifies that the correct IP Filter rule set is installed. The script does the following:
    - Parses /etc/ipf/pfil.ap to determine if any network interfaces are commented out. If some network interfaces are commented out, the script generates a security policy violation message.

- Reviews the existing /etc/ip/ipf.conf file on the system to see if it is the same as the keyword-specific driver. If it is not, the script generates a security policy violation message.
- Verifies that the network/ipfilter service is enabled. If it is not, the script generates a security policy violation.
- enable-password-history.aud: This script verifies the correct configuration of password history on the system. The script checks the /etc/default/passwd file to determine if a HISTORY value is specified:
  - If a HISTORY value is specified in the /etc/default/ passwd file, the script checks it against the value in the JASS_PASS_HISTORY environment variable to see if it is correct.
  - If the HISTORY value is not correct as specified in the JASS_PASS_HISTORY environment variable, the script corrects the value.
  - If the HISTORY value is not set properly, the script corrects the value and issues an audit security violation.
- enable-priv-nfs-ports.aud: This script determines if the NFS service is configured to accept only client communication that originates from a port in the privileged range below 1024. A failure is indicated if the NFS service is not configured properly.
- enable-process-accounting.aud: This script determines if the processing accounting software is installed, enabled, or running on the system. A failure is indicated if this is not true.
- enable-rfc1948.aud: This script determines if the system is configured to use RFC 1948 for its TCP sequence number generation. This script checks both the stored configuration and the actual runtime setting. A failure is displayed if the system is not configured to use RFC 1948-compliant TCP sequence number generation.
- enable-stack-protection.aud: This script determines if the noexec_user_stack and noexec_user_stack_log options are set in the /etc/system file to enable stack protections and exception logging. If these options are not enabled, a failure is reported.
- enable-tcpwrappers.aud: This script determines if TCP wrappers are not installed or configured using the hosts.allow/deny templates included with the Solaris Security Toolkit software or enabled by using the ENABLE_TCPWRAPPERS variable. A failure is reported if the system is not using TCP wrappers.

Install audit scripts

The following install audit scripts are described in this section:

- install-at-allow.aud
- install-fix-modes.aud
- install-ftpusers.aud
- install-jass.aud

Audit Techniques and Collecting Evidence   77

- install-loginlog.aud
- install-md5.aud
- install-nddconfig.aud
- install-newaliases.aud
- install-openssh.aud
- install-recommended-patches.aud
- install-sadmind-options.aud
- install-security-mode.aud
- install-shells.aud
- install-strong-permissions.aud
- install-sulog.aud
- install-templates.aud

Here is a detailed description of performing the task of installing audit scripts:

- install-at-allow.aud: This script determines if a user name is listed in the JASS_AT_ALLOW variable and does not exist in the /etc/ cron.d/at.allow file. The list of user names defined by JASS_AT_ALLOW is empty by default. To pass this check, each user name must exist in both the /etc/passwd file and the /etc/cron.d/at.allow file. Furthermore, a user name should not be in the /etc/cron.d/at.deny file. A failure is displayed if a user name is not listed in both files.
- install-fix-modes.aud: This script determines if the Fix Modes program was installed and run on the system. It indicates a failure if the software is not installed or has not been run. Further, this script uses Fix Modes in debug mode to determine if any additional file system objects should be adjusted.
- install-ftpusers.aud: This script determines if a user name listed in the JASS_FTPUSERS parameter does not exist in the ftpusers file.
- install-jass.aud: This script determines if the Solaris Security Toolkit (SUNWjass) package is installed on the system. A failure is reported if this package is not installed.
- install-loginlog.aud: This script checks for the existence, proper ownership, and permissions for the /var/adm/loginlog file. It indicates a failure if the file does not exist, has invalid permissions, or is not owned by the root account.
- install-md5.aud: This script determines if the MD5 software is installed on the system. A failure is reported if the software is not installed.
- install-nddconfig.aud: This script determines if the nddconfig run-control script files identified in the Sun BluePrints OnLine article, Solaris Operating Environment Network Settings for Security and included with the Solaris Security Toolkit, have been copied to, and their settings made active on, the target system. The script

performs the following checks per object: 1. Tests to ensure that the source and target file types (regular file, symbolic link, or directory) match; 2. Tests to ensure that the source and target file type contents are the same. This script also verifies that the settings defined by the nddconfig script are actually in place on the running system. This script uses its own copy of the nddconfig script in the Solaris Security Toolkit to provide more accurate reporting of results, especially in cases where the script name has changed or where other scripts are used to implement the same effects. This script gives a failure when any of the checks described above are found to be false.

- install-newaliases.aud: This script checks for the existence of the /usr/bin/newaliases program. It indicates a failure if this file does not exist or is not a symbolic link.
- install-openssh.aud: This script determines if the OpenSSH package specified by the script is installed and configured. A failure is reported if the package is not installed.
- install-recommended-patches.aud: This script determines if the patches listed in the Recommended and Security Patch Cluster file are installed on the system. The patch information is collected from JASS_HOME_DIR/Patches directory, based on the Solaris OS version of the system being tested. A failure is displayed if one or more of these patches are not installed. Note that this script indicates success if the version of the patch installed is equal to or greater than the version listed in the patch order file.
- install-sadmind-options.aud: This script determines if the sadmind service exists in the /etc/inet/ inetd.conf file. If it does, this script checks to ensure that options are set to those defined by the JASS_SADMIND_OPTIONS variable. The default setting is -S 2.
- install-security-mode.aud: This script checks the status of the EEPROM security mode. It displays a warning if the mode is not command or full. In addition, this script checks the PROM failed login counter and displays a warning if it is not zero.
- install-shells.aud: This script determines if any shell defined by the JASS_SHELLS parameter is not listed in the shells file. A failure is displayed if any shells listed in JASS_SHELLS are not also listed in the shells file.
- install-strong-permissions.aud: This script determines if any of the modifications recommended by the install-strong-permissions.fin script were not implemented. A failure is displayed if any of these modifications were not made.
- install-sulog.aud: This script checks for the proper ownership and permissions of the /var/adm/sulog file. The script indicates a failure if the file does not exist, has invalid permissions, or is not owned by the root account.

Table 3.1 The shells defined by JASS_SHELLS

/usr/bin/sh	/usr/bin/csh
/usr/bin/ksh	/usr/bin/jsh
/bin/sh	/bin/csh
/bin/ksh	/bin/jsh
/sbin/sh	/sbin/jsh
/bin/bash	/bin/pfcsh
/bin/pfksh	/bin/pfsh
/bin/tcsh	/bin/zsh
/usr/bin/bash	/usr/bin/pfcsh
/usr/bin/pfksh	/usr/bin/pfsh
/usr/bin/tcsh	/usr/bin/zsh

- install-templates.aud: This script determines if the files defined by the JASS_FILES variable were successfully copied to the target system. It indicates a failure if either of the two following checks fail: a test to ensure that the source and target file types match (regular file, symbolic link, or directory) and a test to ensure that their contents are the same.

We have used a file in the install audit script; the details about the file are given in Table 3.1.

Print audit scripts

The following print audit scripts are described in this section:

- print-jass-environment.aud
- print-jumpstart-environment.aud
- print-rhosts.aud
- print-sgid-files.aud
- print-suid-files.aud
- print-unowned-objects.aud
- print-world-writable-objects.aud

These scripts perform the same functions as the print finish scripts, except that they are customized for audit use. The detailed discussion for all is as follows:

- print-jass-environment.aud: This script displays the variables and their content used by the Solaris Security Toolkit. It does not perform any validation or other checks on the content.

- print-jumpstart-environment.aud: This script is for JumpStart mode only. It is used to print out JumpStart environment variable settings. This script does not perform any audit checks.
- print-rhosts.aud: This script displays a notice for any files found with the name of.rhosts or hosts.equiv. Further, this script displays the contents of those files for further inspection.
- print-sgid-files.aud: This script displays a notice for any files that have the set-gid bit set, and it provides a full (long) listing for further review.
- print-suid-files.aud: This script displays a notice for any files that have the set-uid bit set, and it provides a full (long) listing for review.
- print-unowned-objects.aud: This script displays a notice for any files that are not assigned to a valid user and group, and it provides a full (long) listing for further review.
- print-world-writable-objects.aud: This script displays a notice for any matching files that are world-writable, and it provides a full (long) listing for review.

Remove audit script

The following remove audit script is described in this section:

- remove-unneeded-accounts.aud

Details of remove audit script are as follows:
- remove-unneeded-accounts.aud The remove-unneeded-accounts.aud script validates that unused Solaris OS accounts, defined by the JASS_ACCT_REMOVE variable, were removed from the system.

Set audit scripts

The following set audit scripts are described in this section:

- set-banner-dtlogin.aud
- set-banner-ftpd.aud
- set-banner-sendmail.aud
- set-banner-sshd.aud
- set-banner-telnet.aud
- set-flexible-crypt.aud
- set-ftpd-umask.aud
- set-login-retries.aud
- set-power-restrictions.aud
- set-rmmount-nosuid.aud
- set-root-group.aud
- set-root-home-dir.aud
- set-root-password.aud

- set-strict-password-checks.aud
- set-sys-suspend-restrictions.aud
- set-system-umask.aud
- set-term-type.aud
- set-tmpfs-limit.aud
- set-user-password-reqs.aud
- set-user-umask.aud

Here is the detailed discussion about all set audit scripts:

- set-banner-dtlogin.aud: This script verifies that a service banner for the CDE or dtlogin service is defined. This script verifies that the system displays the contents of /etc/motd by listing it in the file template JASS_ROOT_DIR/etc/ dt/config/Xsession.d/0050.
- set-banner-ftpd.aud: This script checks that the FTP service banner matches the value defined by the JASS_BANNER_FTPD variable. It indicates a failure if the service banner does not match. The value of the variable is Authorized Use Only.
- set-banner-sendmail.aud: This script verifies that the sendmail service is configured to display the service banner as defined by the JASS_ BANNER_SENDMAIL environment variable. This banner is displayed to all clients connecting to the sendmail service over the network.
- set-banner-sshd.aud: This script verifies that the Secure Shell service banner is displayed by ensuring that the Secure Shell service displays the contents of / etc/issue to the user prior to authenticating access to the system.
- set-banner-telnet.aud: This script checks that the Telnet service banner matches the value defined by the JASS_BANNER_TELNETD variable. It indicates a failure if the service banner does not match.
- set-flexible-crypt.aud: This script verifies the use of strong passwords by checking that the changes described in "Invalid Cross-Reference Format" for each of the Solaris Security Toolkit drivers have been made correctly.
- set-ftpd-umask.aud: This script checks that the FTP service banner matches the value defined by the JASS_FTPD_UMASK variable. It indicates a failure if the file creation mask value does not match. The value of the variable is 022.
- set-login-retries.aud: This script determines if the login RETRIES parameter is assigned the value defined by the JASS_LOGIN_ RETRIES variable. The variable default is set to 3. A failure is displayed if the variable is not set to the default.
- set-power-restrictions.aud: This script checks the /etc/default/ power file and indicates a failure if the PMCHANGEPERM and CPRCHANGEPERM parameters do not have a hyphen "-" as their values.

- set-rmmount-nosuid.aud: This script determines if the /etc/rmmount .conf file restricts the mounting of a removable Unix File System (UFS) or a High Sierra File System (HSFS) by enforcing the nosuid parameter. A failure is displayed if this restriction is not defined in the /etc/rmmount.conf file.
- set-root-group.aud: This script determines if the root account's primary group is set to the value defined by the JASS_ROOT_GROUP variable. A failure is displayed if it is not defined properly.
- set-root-home-dir.aud: This script checks to see if the root account has a home directory of / in the /etc/passwd file:
  - If the home directory is /, the script generates an audit error.
  - If the home directory is /root, the script checks the following:
    - Directory ownership should be root:root o Directory permissions should be 0700.
    - Dot files (/.cshrc, /.profile, /llogin, /.ssh) are moved from /to / root.
    - Dot file permissions should all be 0700.
  - If the home directory is neither / nor /root, the script generates a warning, but not an audit error.
- set-root-password.aud: This script checks the password of the root account. It indicates a failure if the value is the same as that of the JASS_ROOT_PASSWORD variable. This check is done to encourage users to change the root password from the value defined by JASS_ROOT_PASSWORD as soon as possible.
- set-strict-password-checks.aud: This script verifies that the correct values for the various password checks are defined correctly in the /etc/default/passwd file.
- set-sys-suspend-restrictions.aud: This script checks the /etc/default/sys-suspend file. It indicates a failure if the PERMS parameter does not have a hyphen "-" as its value.
- set-system-umask.aud: This script determines if the system's default file creation mask is set to the value defined by the JASS_UMASK variable. The default value is set to 022. A failure is displayed if the variable is not properly defined.
- set-term-type.aud: This script determines if the /etc/profile and the /etc/login files set the default terminal type to vt100. A failure is displayed if the default terminal type is not defined properly. This script is provided as a convenience only, and a failure does not impact the security of a system.
- set-tmpfs-limit.aud: This script determines if any tmpfs file systems are defined in the /etc/vfstab file without their size being limited to the JASS_TMPFS_SIZE variable, which is set to a default of 512 megabytes. A failure is reported if the tmpfs file system size does not comply with the JASS_TMPFS_SIZE value.

- set-user-password-reqs.aud: This script reviews the password policy settings on the system as defined previously. It indicates an error if the values do not match the following default values defined by the Solaris Security Toolkit:
  - MINWEEKS – 1
  - MAXWEEKS – 8
  - WARNWEEKS – 1
  - PASSLENGTH – 8

  The default values are contained in the following environment variables:
  - JASS_AGING_MINWEEKS
  - JASS_AGING_MAXWEEKS
  - JASS_AGING_WARNWEEKS
  - JASS_PASS_LENGTH
- set-user-umask.aud: This script determines if any of the following files do not set the umask parameter to the value defined by the JASS_UMASK variable, whose default value is set to 022. A failure is displayed if these files do not set the unmask parameter appropriately.

Update audit scripts

The following update audit scripts are described in this section:

- update-at-deny.aud
- update-cron-allow.aud
- pdate-cron-deny.aud
- update-cron-log-size.aud
- update-inetd-conf.aud

The detailed description of all updated audit scripts is as follows:

- update-at-deny.aud: This script determines if a user account is listed in the JASS_AT_DENY variable and is not listed in the /etc/cron.d/at.deny file.
- update-cron-allow.aud: This script determines if a user account is listed in the JASS_CRON_ALLOW variable and not in /etc/cron.d/cron.allow file. By default, the value is only the root user. A failure is displayed if this check fails.
- update-cron-deny.aud: This script determines if a user account is listed in the JASS_CRON_DENY variable and not in the /etc/cron.d/cron.deny file.
- update-cron-log-size.aud: This script determines if the cron facility is configured to increase its default size limit for log files. The check method is based on the version of the Solaris OS and the value of

Table 3.2 Sample output of JASS_SVCS_DISABLE

100068	100083	100087	100134	100146	100147
100150	100155	100166	100221	100229	100230
100232	100234	100235	100242	100424	300326
536870916	chargen	comsat	daytime	discard	dtspc
echo	eklogin	exec	finger	fs	ftp
kerbd	klogin	kshell	login	name	netstat

the JASS_CRON_LOG_SIZE variable. The size limit defined by the JASS_CRON_LOG_SIZE variable is 20,480 kilobytes. A failure is displayed if the size limitation is not correct.

- update-inetd-conf.aud: This script determines if any of the services listed in the JASS_SVCS_DISABLE variable are disabled in /etc/inetd.conf. This script also checks to ensure that services listed in the JASS_SVCS_ENABLE variable are enabled in the /etc/inetd.conf file. If a service is listed in both variables, then the service is left enabled by the JASS_SVCS_ENABLE variable. A failure is displayed if any of these checks fail.

Sample output of JASS_SVCS_DISABLE audit scripts mentioned in Table 3.2 provides more intuitions of results.

Using product-specific audit scripts

Table 3.3 lists product-specific audit scripts for specific Sun products. These scripts are in the Audit directory. New audit scripts are released periodically for new and updated Sun products. For the latest list of scripts, refer to the Security Web site: http://www.sun.com/security/jass.

- suncluster3x-set-nsswitch-conf.aud: This script determines if the /etc/nsswitch.conf file lists the cluster keyword as the first source for the host's database. A failure is displayed if this is not true. For more information, refer to the Sun BluePrints OnLine article titled "Securing Sun Cluster 3.x Software."

Table 3.3 Product-specific audit scripts for specific Sun products

Product	Driver Name
Sun Cluster 3.x software	suncluster3x-set-nsswitch- - conf.aud
Sun Fire high-end systems domains	s15k-static-arp.aud
Sun Fire high-end systems system controllers	s15k-static-arp.aud s15k-exclude-domains.aud s15k-sms-secure-failover.aud

- s15k-static-arp.aud: For System Management Services (SMS) versions 1.2 through 1.4.1, this script verifies that the static ARP configuration files are installed on Sun Fire high-end systems system controllers (SCs) and domains. For system controllers, the file is /etc/sms_sc_arp. For domains, the file is /etc/sms_domain_arp. This script checks that all existing domains have Ethernet addresses as listed in the SC static ARP startup script and corresponding data file.
- s15k-sms-secure-failover.aud: For SMS versions 1.2 through 1.4.1, this script determines if the Sun Fire high-end systems system controller is configured based on the recommendations in the Sun BluePrints OnLine article titled "Securing the Sun Fire 12K and 15K System Controller." It indicates a failure if any of the services listed in the SMS_SVCS_DISABLE variable are enabled in /etc/inet/ inetd.conf.

3.3 AUDIT STAGES

Stages of an audit and how an audit is conducted can differ depending on the size of the corporation and the complexity of the case. However, an audit usually has four main stages; also refer to Figure 3.1 audit stages:

- The first stage is the planning stage. In this stage, a corporation engages with the auditing firm to establish details, such as the level of engagement, procedures, and objectives.
- The second stage is the internal controls stage. In this stage, auditors gather financial records and any other information necessary to conduct their audits. The information is necessary to evaluate the accuracy of the financial statements.
- The third stage is the testing stage. In this stage, auditors examine the accuracy of the financial statements using various tests. It may involve verifying transactions, overseeing procedures, or requesting more information.

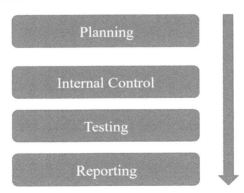

Figure 3.1 Audit stages.

- The fourth stage is the reporting stage. After completing all the tests, the auditors prepare a report that expresses an opinion on the accuracy of the financial statements.

Levels of audit engagement

Many companies choose to engage with internal and external auditors in the preparation of their year-end financial statements. However, the depth of the auditor's investigation may vary depending on the type of engagement and the assertion level required.

In a full audit engagement, the auditor conducts a complete and thorough investigation of the financial statements, including verifications of income sources and operating expenses. For example, the auditor may compare reported account receivables with receipts from actual customer orders.

At the end of the engagement, the auditor will provide an opinion on the accuracy of the financial statements. A full audit engagement also provides investors, regulators, and other stakeholders with confidence in a corporation's financial position.

In a review engagement, an auditor only conducts limited examinations to ensure the plausibility of the financial statements. In contrast with an audit, the review engagement only assures that the financial statements are fairly stated, and no further examinations are conducted to verify the accuracy of the statements. Therefore, a review engagement does not provide the same level of confidence in the accuracy of the financial reporting relative to an audit.

In a notice to reader engagement, the role of the auditor is solely to help a company compile its financial information into presentable financial statements. No further examinations are performed, and no opinions are expressed on the accuracy of the financial reporting. Notice to reader engagements is typically only utilized by small corporations without any obligations to external stakeholders.

3.4 AUDIT TECHNIQUES

Audit techniques stand for the methods that are adopted by an auditor to obtain evidence. In this reference, there are various techniques or ways, but we are going to focus on a few only. The detailed description of all techniques is as follows:

Inspection

Documents and records
  While verifying various transactions, the auditor examines the supporting documents and records. This technique is otherwise called vouching. The purpose of examining the documents and records is to confirm the authenticity

(genuineness) of the transaction, to find whether the transactions and the supporting document are appropriate, to ensure whether the transactions are authorized, and to ensure whether the classification of the transaction is proper.

The auditor goes through the accounting records and documents and if he comes across any unusual transactions, he verifies the same thoroughly. This is called scanning of records, which requires experience and expertise. How an auditor can rely on the documents depends on the origin (source) of the documents and the efficiency of the internal control system in operation? Documents which have their origin in the hands of the third parties and held by third parties are more reliable than the documents which have their origin in the organization itself and held by the organization. One can classify the documents into four major categories according to their origin and availability.

- Documents which have their origin in the hands of the third party and held by them – Most reliable evidence.
- Documents which have their origin in the hands of the third party and held by the organization – More reliable.
- Documents which have their origin in the hands of the organization and held by the third party – Reliable.
- Documents which have the origin in the hands of the organization and held by the organization – Reliable only if the internal control is effective.

Physical verification

If an item can be measured in physical terms, the same may be verified for quantity and quality (if possible). By physical examination, the auditor ensures the availability of the item. However, the ownership of the items cannot be verified through this method.

Observation

The auditor observes a particular procedure being carried out by the organization. Examples are observation of the internal control measures that are adopted in transactions involving cash, procedures followed on receipt or issue of material, etc. The auditor makes his observations to evaluate the efficiency and effectiveness of the system followed by the organization.

Inquiry and confirmation

Inquiry: Seeking information from persons belonging to the organization or from outside organization is called inquiry.

Confirmation: Confirming the information available with the records of the organization or with the persons mostly from outside the organization through an inquiry is confirmation.

Inquiry and confirmation can take place either orally or in writing. The best example for inquiry and confirmation is confirming the balances of debtors shown in the accounting records with the debtors of the organization.

Computation

An auditor makes appropriate calculations and verifies the accuracy of the accounting records. For example, the auditor computes the depreciation to be charged for the year, by taking into consideration the value of the asset (cost), the date of purchase, the rate of depreciation, etc., to verify the accuracy of the depreciation charged by the organization. The auditor also traces a particular transaction from the origin to check the book keeping procedure.

Analytical procedures

The purpose of the analysis is to ensure consistency of accounting methods and also to evaluate the efficiency of the management by comparing the results over several years. The several analytical procedures are reconciliation, ratio analysis, and variance analysis.

The auditor also applies analytical procedures to help the management in decision making. Such analytical techniques are marginal costing, standard costing, etc.

The auditor studies the nature of the business and also the prevailing circumstances and selects the techniques to be applied. While conducting the audit, he may change his technique according to the changes observed in the circumstances. The suitable audit techniques adopted by the auditor help him to carry on the audit efficiently.

3.5 COLLECTING EVIDENCE THROUGH QUESTIONS

Audit evidence is all the information used by the auditor in arriving at the conclusions on which the audit opinion is based and includes the information contained in the accounting records underlying the financial statements and other information. Auditors are not expected to examine all information that may exist.

Following are various ways of collecting evidences:

- Inspection.
- Observation.
- External configuration.
- Documentation.
- Recalculation.

- Re-performance.
- Analytical procedures.

Inquiry

Audit evidence, which is cumulative in nature, includes audit evidence obtained from audit procedures performed during the course of the audit and may include audit evidence obtained from other sources, such as previous audits and a firm's quality control procedures for client acceptance and continuance.

Evidence includes information that is highly persuasive, such as the auditor's count of marketable securities, and less persuasive information, such as responses to questions of client employees.

The use of evidence is not unique to auditors. Evidence is also used extensively by scientists, lawyers, and historians. An auditor must gather sufficient and appropriate audit evidence and test them to make a judgment of opinion.

In gathering evidence to support his assertions, the auditor is often confronted with two issues: What evidence will be relevant to assess an assertion with greater reliability? And how much evidence is to be obtained?

Sufficient appropriate audit evidence

The auditor should design and perform audit procedures that are appropriate in the circumstance for the purpose of obtaining sufficient appropriate audit evidence. Sufficiency is the measure of the quantity of audit evidence.

Appropriateness is the measure of the quality of audit evidence, that is, its relevance and its reliability in providing support for, or detecting misstatements in, the classes of transactions, account balances, and disclosures and related assertions.

The auditor should consider the sufficiency and appropriateness of audit evidence to be obtained when assessing risks and designing further audit procedures.

The quantity of audit evidence needed is affected by the risk of misstatement (the greater the risk, the more audit evidence is likely to be required) and also by the quality of such audit evidence (the higher the quality, the less the audit evidence that may be required).

Accordingly, the sufficiency and appropriateness of audit evidence are interrelated. However, merely obtaining more audit evidence may not compensate if it is of lower quality.

Ways of collecting audit evidence

Inspection

The inspection involves examining records or documents, whether internal or external, in paper form, electronic form, or other media, or a physical examination of an asset.

Inspection of records and documents provides audit evidence of varying degrees of reliability, depending on their nature and source and, in the case of internal records and documents, on the effectiveness of the controls over their production.

An example of inspection used as a test of controls is the inspection of records for evidence of authorization.

Observation

Observation consists of looking at a process or procedure being performed by others, for example, the auditor's observation of inventory counting by the entity's personnel or of the performance of control activities.

Observation provides audit evidence about the performance of a process or procedure but is limited to the point in time at which the observation takes place and by the fact that the act of being observed may affect how the process or procedure is performed.

External confirmation

An external confirmation represents audit evidence obtained by the auditor as a direct written response to the auditor from a third party (the confirming party), in paper form, or by electronic or another medium.

External confirmation procedures frequently are relevant when addressing assertions associated with certain account balances and their elements.

Documentation

Documentation is the auditor's examination of the client's documents and records to substantiate the information that is or should be included in the financial statements.

The documents examined by the auditor are the records used by the client to provide information for conducting its business in an organized manner.

Because each transaction in the client's organization is normally supported by at least one document, there is a large volume of this type of evidence available.

Recalculation

Recalculation consists of checking the mathematical accuracy of documents or records. Recalculation may be performed manually or electronically.

Re-performance

Re-performance involves the auditor's independent execution of procedures or controls that were originally performed as part of the entity's internal control.

Analytical procedures

Analytical procedures consist of evaluations of financial information through analysis of plausible relationships among both financial and non-financial data.

Analytical procedures also encompass such investigation as is necessary for identified fluctuations or relationships that are inconsistent with other relevant information or that differ from expected values by a significant amount.

Inquiry

Inquiry consists of seeking information from knowledgeable persons, both financial and non-financial, within the entity or outside the entity. An inquiry is used extensively throughout the audit in addition to other audit procedures. Inquiries may range from formal written inquiries to informal oral inquiries.

Evaluating responses to inquiries is an integral part of the inquiry process.

3.6 OBSERVATION

Observation consists of looking at a process or procedure being performed by others, for example, the auditor's observation of inventory counting by the entity's personnel or of the performance of control activities.

Observation provides audit evidence about the performance of a process or procedure but is limited to the point in time at which the observation takes place and by the fact that the act of being observed may affect how the process or procedure is performed.

3.7 REPORTING TO AUDIT FINDING

If you are conducting an internal audit, you may wonder how to report audit findings. Typically, internal and external audit findings are reported in writing as well as delivered verbally to stakeholders. For external and internal audits, there is a general structure that most written reports follow. For compliance audits, the overseeing organization may have specific requirements for what needs to be included in the written report. A written audit report should be written concisely and, in a way, that's easily understood by the reader. It should also include evidence for any audit findings.

In general, an audit report has three sections: an introduction, a section which describes the scope of the audit, and the auditor's opinion, which describes the audit findings. The introduction states the auditor's responsibilities and your business responsibilities regarding the audit. It would typically also include the names of the auditor or auditors and the dates of the audit. The scope section describes the auditing process. It states the areas that were audited, who completed the audit, and when and what criteria were used to perform the audit. If it is an external audit, for example, it would state which governing body's standards were being used to measure the results of the audit. For an internal audit, it would refer to the company standards and policies that were being used. The scope section also describes exactly what the auditor did. The auditor would include what financial statements she reviewed and what tests she performed.

The auditor's opinion is the final section of the report. This is where the auditor states what she found and whether your business conforms to the criteria of the audit. Depending on the type of audit, the auditor may also include recommendations for improving or solving issues that were found during the audit.

After writing her report, the auditor will typically present her findings to stakeholders within the company. In a large nonprofit organization, for example, she would present her findings to an audit committee, which oversees the auditing process. The committee would discuss the audit findings with the auditor and ask clarifying questions before they present the audit report to their board of directors.

If you are conducting an internal audit, the best approach for how to write audit findings and recommendations is to clearly state what you did, when you did it, and what you found. Your report doesn't need to be lengthy. It does need to be clear and compelling so that your business managers and executives will take action based on your results, especially if you find areas that need improvement.

Different types of audit findings

A thorough audit may uncover areas of weakness that need to be improved. In the worst-case scenario, it may uncover fraud or mismanagement that you need to appropriately report to authorities and address. External audits typically report the audit findings as one of the following: an unqualified or clean opinion, a qualified opinion, an adverse opinion, or a disclaimer of opinion.

An unqualified opinion is a best-case scenario if your business is going through an external audit. It means that the auditor was able to complete the audit and that your business was in compliance with the criteria of the audit. For example, in a financial statement audit, an unqualified opinion would mean that the statements conform to generally accepted accounting principles.

A qualified opinion means that there was an issue with the audit. The auditor may not have been given access to all the information and documents he needed, for example, or there may not have been compliance with the accepted standards for the area being audited. In a financial statement audit, this may mean that the auditor found one or more areas where generally accepted accounting principles were not being followed. Although this isn't the ideal result for your business, it can be quickly and easily addressed.

An adverse opinion is much more serious. It indicates that the auditor found a misrepresentation or misstatement in the area being audited. In the case of a financial statement audit, an adverse opinion means that the auditor found that your company's financial statements don't align with generally accepted accounting practices. This result is relatively rare, and it can have a negative impact on publicly traded companies. Some companies have seen a fall in their stock prices, for example, after an adverse opinion has been issued.

A disclaimer of opinion means that the auditor was unable to complete the audit. This may be because the financial statements weren't available or that the auditor wasn't given full access to the information he needed. It may also mean that your company's management was not cooperative with the auditor. It sometimes also indicates that there was a conflict of interest on the part of the auditor. For example, he may have a financial interest in the company being audited.

Although a disclaimer of opinion isn't the best scenario for you or your company, it isn't the worst, either. It essentially means that there is no opinion and that the audit should be completed at a future time.

Respond to audit findings

If your business has recently completed an internal or external audit, it's critical to respond to the audit findings. The first step is to carefully review the audit report with the auditors. Ask the auditors questions to clarify the findings and their experience when working with your company. You may ask about how cooperative your staff members were with the auditors, for example. You might also ask if there were any changes to the original audit plan or if the auditor experienced any difficulties during the audit process.

If the auditor found areas that need improvement, ask about specifics. Make sure you explain what the issue is and what document or test uncovered the issue. You may also ask if the auditor has any specific recommendations for resolving the area. Even if no issues were found, you can still take advantage of your auditor's knowledge and experience. You can ask how your organization compares to other similar organizations, for example. You can also ask for general recommendations for improving your accounting procedures or reporting practices.

In the event of an adverse or qualified opinion, you should consider providing a formal, written response. You may be required to respond in writing, depending on the policies of your company or the governing body overseeing your audit. Your response should directly address each issue raised in the report and discuss your plan.

3.8 AUDIT TEAM MEETING

Importance of opening meetings

Many conformity assessment bodies (CABs) do not conduct effective opening meetings. These are extremely important and a valuable part of the audit process for obvious reasons. We have witnessed many auditors in different parts of the world conducting management system audits and not paying much attention to this important element. They simply rush into the auditing of the management system(s) without setting the tone for the audit in a positive manner.

Below is a breakdown of how each on-site management system audit should be started (opening meeting).

Opening meeting

The opening meeting is one of the most important events in the history of our dealings with new and existing clients, and therefore, it must be handled with the utmost precision in terms of its conduct. It introduces, possibly for the first time, real live representatives of our company and can set the scene for the audit and all future dealings with the client. It is a great forum to create a relaxed atmosphere and to put the audit organization (client) at ease.

The following is the minimum agenda that we cover at the opening meeting. We aim to make sure that the client is extremely clear about each thing we table at every opening meeting. There is nothing worse than the auditees not understanding what is happening during each part of the audit. We encourage our clients to question everything that they do not fully understand.

Introduction

We always introduce the audit team members, and we sometimes lighten things up by saying something like "You have the pleasure of the company of my colleague and myself for the next 8 hours (two days, or whatever), I hope it will be a pleasure and not the reverse!" We try to put the client at ease so they get value from the audit event, no-one likes stressful audits so we try so hard to ensure the client is not stressed.

Confirm the scope and objectives of the assessment

We either confirm that the scope is unchanged or we ascertain exactly what the new scope wording is that is required by the client. We must remember that the scope cannot be developed into something that entails a new industry classification or the audit team may not be correctly qualified to then proceed with the audit. The new scope should be recorded on the Opening Meeting Checklist or it can be left to the closing meeting if the company wishes to have more time to discuss the exact wording among themselves.

We also need to explain the exact objective of each audit visit. The objective is usually to check the organization's level of compliance against the chosen management system standard(s).

Confirm communications, resources, and escorts

We always confirm who our lines of communication will be, who is available to help us to ensure that the audit runs smoothly, and who will be showing us around and introducing us to the key audit personnel.

Current number of employees

We always check the numbers of employees for each company we audit at the opening meeting to ensure that we are tracking the size of the organization.

Confirm auditor confidentiality

We should point out that everything that the audit team sees during the audit will remain confidential (even though as one MD said to one of our team leaders, if you tell me what our opposition does in their business, we will keep it confidential!!). We must also explain how we all sign confidentiality agreements that are legally binding to ensure confidentiality is maintained at all times.

Explain the audit program and the reporting process for deficiencies

We feel it is important to run through the audit plan at the opening meeting to ensure that there are no surprises for the client, and we give an overview of what we will be looking at throughout the course of the audit event.

We should point out that if there are only minor corrective action requests (CARs) at the end of the audit, then the client has to do nothing but commit to correcting the deficiencies and giving their responses as required. If there are any major CARs, then the client has to either complete the corrective action and have it re-audited to close out the CAR or do enough of the corrective action such that if the remaining problem were on its own, it would be graded as a minor and as such the major can now be downgraded to minor status. Registration cannot proceed with open major CARs or no responses to minor CARs.

Confirm time and place for closing meeting

We explain that the schedule is flexible and if managers or their employees are required to carry out certain tasks during the course of the audit, then the audit team will try to accommodate this as much as possible.

It is essential that the closing meeting time is confirmed, this should have been printed on the assessment schedule sent to the client in advance of the audit. Under no circumstances should this meeting be late; if there are extenuating circumstances over which the audit team had no control, then the management team should be informed as early as possible of the new meeting time.

Appeals process

We always explain that if the client does not agree with any CARs, then there is an appeals process that is outlined in the certification criteria that all clients are issued with.

Audit team safety induction

We always ask about the safety of the audit team and whether there are any significant hazards we may encounter. We request an induction if it has not been offered by the client.

Each client will normally have this covered without needing prompting from us.

3.9 NONCONFORMITIES AND OBSERVATION

The term "nonconformity" inspires fear in many people. It shouldn't. A nonconformity is simply an opportunity for the management system to improve. It should not be viewed as an indictment of any person or group but rather as a factual statement that drives improvement. Before we get any further, it's helpful to provide a clear definition of what a nonconformity is. A nonconformity is the failure to meet a requirement.

It's a short definition, but it packs a lot of power. The first thing you should notice is the prerequisite. You need a requirement before you can ever have a nonconformity. When we write a nonconformity, we always write the requirement first. It sets the tone for everything else. If you can't find a requirement for a particular situation, then categorically you can't have a nonconformity. You might have a concern, observation, remark, or opportunity, but it's not a nonconformity unless it's clearly tied to a requirement.

The second half of a nonconformity is the objective evidence. The evidence states exactly what the auditor saw, heard, read, or experienced that contradicted the requirement. The objective evidence is factual and traceable, but it is stated as concisely as possible. We can't write a book detailing every minute thing that happened. A good auditor simply cites the evidence that fails to meet the requirement.

A child could write a clear nonconformity. It's a simple one-two process. This is how the two halves of a nonconformity statement fit together:

Requirement: The company committed itself to do XYZ. The commitment is a fact, evidenced by its presence in a procedure, plan, policy, specification, contract, work instruction, standard, or statement.

Evidence: The company failed to do XYZ. The failure is a fact, based on evidence such as records, observations, documents, or interviews.

No opinions are present in the nonconformity, just cold, hard facts. It's hard to argue with facts. It also makes the audit go much smoother. Sure, facts may remove a degree of creativity that auditors exercised, but creativity is better expressed in other ways.

Writing effective nonconformities is one of the most fundamental auditing skills. Despite its fundamental nature, it is a skill that even the most experienced auditors struggle with. Here are a few keys that will help you write nonconformities as well as anybody out there.

Match the requirement with concise evidence: This is the single most important key. State the requirement and then provide the evidence that shows that the requirement was not met.

Write in complete sentences: This is what your eighth-grade English teacher would have insisted on, and good auditing insists on it, also. Complete sentences, in both the requirement and the evidence, provides the customer of the audit with a complete product. This complete product is more likely to be understood and thus more likely to be acted upon.

Include all applicable identifiers (what, who, when, where): It is critical that your evidence is fully traceable. This is achieved by including all identifiers: what the nonconformity was, who was involved, when it happened, where was it located, and how much was involved. This builds credibility in the evidence and allows the auditee to know exactly what needs to be done to remedy the situation. The only identifier that is not clearly stated is people's names. We use job titles in the evidence because we want to remind everybody involved that the audit is about the process, not people.

Use an economy of words: Writing a nonconformity is a balancing act. We need to include all identifiers, but we need to use an economy of words. As Shakespeare said, "Brevity is the soul of wit". Not only is brevity the soul of wit, but it helps drive understanding. Ironically, including more words rarely increases anybody's understanding of what you're trying to communicate. Keep your nonconformity statement nice and tight. If you can remove a few words, while still communicating the essential message, then by all means do it.

State the facts, not your opinions: The audit is about facts. There is no need to editorialize as part of the evidence. This happens when the auditor tries to explain why the nonconformity is harmful or what the effects could be. This is not necessary. Just state the evidence and no more.

Example of a well-written nonconformity

Requirement: That employees must wear white gloves when handling finished product.

We state exactly where the requirement comes from. In this case, it's a procedure written by the organization being audited. We provide the procedure number, revision, and even the section. Some auditors also include the procedure name, and this can add value too. Once we say where the requirement comes from, we state exactly what the requirement is. We don't paraphrase it or get creative; we repeat the requirement word for word.

Evidence: The auditor observed two employees in the warehouse handling finished product without white gloves.

The language in the evidence mimics the language in the requirement. The organization committed to wearing white gloves when handling finished products, but the auditor observed two people not wearing white gloves. Cut and dry.

Here is an example of a poorly written nonconformity:

Requirement: All products must be identified.

There is no traceability to this requirement. Where did it come from? There is a similar requirement in ISO 9001:2015, but that source is not referenced. Always cite where the requirement is coming from, and never paraphrase it. It is permissible to add an ellipsis if the requirements are taken from a long passage and the entire section is not needed.

Evidence: Returned goods were missing the nonconforming materials tags, which greatly increases the chance of accidentally shipping bad material.

This evidence has a lot of problems, most of which can be summarized by saying the evidence is not traceable. Firstly, there's no indication of where the returned goods were located. The returned goods are also not identified in any unique way. The quantity of returned goods is also not indicated. Finally, the auditor included a lot of editorial opinions at the end of the evidence. This opinion is not needed and only serves to inflame the auditee. Stick to the facts when you write evidence and make sure the facts are fully traceable.

Auditors are held to a higher standard

It's tempting to just say "Who cares?" when it comes to writing clear and correct nonconformity statements. After all, the rest of the world writes in a very informal manner. Grammar, correct spelling, and proper usage all seem optional these days. As long as the auditee understands what I'm talking about, it's okay, right? Wrong answer. Auditors must produce a polished and professional product. Everything you write, say, and do is being scrutinized constantly by the auditee. Not in an attempt to disparage you, but simply to ensure that audit is credible. The output of the audit that gets the most attention is nonconformity write-ups. Even though nonconformities are not aimed at any particular person in an organization, auditees are still keenly aware that they represent failings of a sort. A requirement was established, but we weren't able to meet it. Because of the high profile of nonconformities, they need to be written very well.

If you're auditing with another person or as part of a team, make sure to have someone else review your nonconformity write-up. This is usually the role of the lead auditor, but really any set of trained eyes is helpful. Any rework that is necessary should be performed by the auditor who originally wrote the nonconformity. Recognize also that sometimes the answer is not reworking the nonconformity; it's ripping it up. These are the conditions that usually result in a nonconformity being removed and not being provided as part of the final report:

- Requirement does not actually exist.
- Requirement exists, but it's outside the scope of the audit.
- Evidence is not traceable and it's too late in the audit to gather the details.
- Evidence relies on hearsay, rumors, or second-/third-hand information.
- Auditee has presented evidence (after the original interview) that shows the requirement was met.

This is a good time to address findings that fall outside the scope of the audit. They could be in a department/process that is not officially being audited, or they could be against a standard that is not being used in the present audit. For instance, we see something happening in the warehouse that constitutes a nonconformity, but the warehouse is not part of the audit. Or the auditors observe a safety violation, but the audit is against ISO 9001 and safety is not included. In these cases, it is usually of verbally reporting the finding to your contact within the auditee organization. Verbally report it, mention that it is officially outside the scope, and ask him/her if they would like you to include it in the report. If they say "no," then you don't include it. The scope of the audit is almost a contract, and you don't intentionally audit outside the scope or report findings that are outside the scope. It's perfectly fine to make a note of the issue and follow up the next time there is an audit that includes that topic.

3.10 CORRECTIVE AND PREVENTIVE ACTIONS

Quality management is a broad and deliberate discipline. Of the many concepts it covers, you'll probably have heard quite a bit about corrective action and preventive action. What do these terms mean, and why do they matter? We can go ahead and answer the latter first: A fuller understanding of quality management processes is only possible when you understand these terms. This article discusses what corrective and preventive actions refer to.

An in-depth look at corrective and preventive action

Corrective action and preventive action operate in tandem. These concepts are collectively known as CAPA. Both practices are concepts within HACCP/HARPC (Hazard Analysis and Critical Control Points/Hazard Analysis) and Risk-Based Preventive Controls/and Good Manufacturing Practices (GMP). It also belongs to various ISO business standards.

The focus of CAPA is the systematic investigation of the root causes of identified risks or problems in a bid to ensure they don't occur (preventive) or recur (corrective).

Appropriate CAPA documentation is a requirement of the ISO 9000. It's a similar case with related standards such as the AS9100.

CAPA is a body of regulation- or law-required actions that a company undertakes in documentation, manufacturing, systems, or procedures to eliminate or resolve recurring nonconformance.

Corrective action

Project managers view corrective action as a deliberate activity to realign the performance of project work with a project management plan.

Corrective action is a quality management procedure that involves a sequence of actions an individual or organization performs to rectify a behavior or process. Corrective action is necessary when there is the danger of a production error or a deviation from the original goal or plan.

In simpler terms, the above definition of corrective action is a future response to repairing a defect. It ensures that the error never occurs again.

Suppose we find some defective components and corrected them, for instance. Our goal will be to ensure it doesn't happen again, so we go after the root cause of the problem and develop a solution. Managers will then feature this solution in our processes to ensure the defects do not occur again.

So corrective action enables you to resolve the root cause of the problem while ensuring a repeat of the deviation doesn't happen again. Corrective action is reactive.

Corrective actions aim to address internal audit issues, customer complaints, and undesirable product nonconformance. As Statistical Process Control (SPC) identifies, they may also deal with unstable or unfriendly trends in process and product monitoring.

What's the scope of corrective action?

A corrective action plan aims to identify a problem and use available means and resources to address any symptoms.

The primary goal of corrective action is to identify the source or root cause of a problem and take appropriate steps. Corrective action will typically address critical issues, health dangers, safety concerns, and supply concerns. It also takes care of customer requests for change in fit, form, function, and other recurring problems.

Citing human resources as an example scenario, corrective action helps communicate with employees on performance expectations and acceptable behavior. Corrective actions come into play as soon as performance measurements and coaching do not work.

A corrective action plan aims to identify a problem and use available means and resources to address any symptoms.

Benefits of corrective action

Corrective actions are beneficial in the following ways:

- They specify the steps to solve issues.
- They ensure that activities are transparent.
- They give power to teams.
- They provide a basis for future development and needs.
- They eliminate the need for a problem-solving wheel.

Issues of corrective action

Even with its impressive benefits, we can expect corrective actions to have some drawbacks. These happen if there's poor implementation of the corrective action such that it becomes a merely bureaucratic routine. In such cases, action requests may receive treatment for minor incidents.

The other issue involves focusing on the symptoms instead of the fundamental causes.

Corrective Action Request (CAR)

A corrective action request (CAR) is a feature of manufacturing or production policies. It may be the result of an audit, customer complaints, or a production line occurrence.

As a formal request, CAR aims to remove all sources of nonconformity. Manufacturing nonconformity often stems from the product or production process.

Corrective action requests may address various levels of concern.

Preventive action

Preventive action is a measure an organization takes to forestall any nonconformity to an organization's primary intentions.

Project management professionals describe preventive action as an intentional activity to ensure the project alignment plan and the future performance of project work are in sync.

Quality management preventive action ensures the prevention of any future defects.

If we chose to begin a production process, we may suspect that defects would crop up during production. We may review the processes and work to prevent any future defects.

Preventive action is proactive. In contrast to corrective action, preventive action anticipates a problem and takes appropriate measures to prevent them from happening.

Preventive action aims to identify possible sources of nonconformity.

Quality management preventive action ensures the prevention of any future defects.

What's the scope of preventive action?

Preventive action usually involves predicting possible problems and crafting fitting plans to mitigate them. Preventive plans help organizations uncover possible deficiencies in their operations before developing measures to prevent them.

What does this identification process involve? We'll find elements such as analysis, internal audits, customer feedback review, full employee and worker participation at every level.

Preventive action involves the following:

- Stages of investigations.
- Reviewing analysis.
- Taking relevant actions.
- Reviewing results.
- Taking more action as needed.

It's easy to see how these processes fit snugly into the Deming–Shewhart cycle's Plan–Do–Check–Act (PDCA).

How does corrective action differ from preventive action?

In exploring how corrective actions diverge from preventive actions, definitions are a good place to begin. Corrective actions consist of plans for identifying the root causes of a problem and plotting solutions for the outcomes. Here are other points of difference:

- Preventive actions comprise actions for identifying potential risks that could impact operations and create plans to mitigate them.
- If we can consider the point of use, organizations apply preventive actions before problems arise. Therefore, preventive actions are a way to resolve consequences and minimize additional risks.
- Some corrective actions in manufacturing involve the recall of substandard products after their launch on the market. In human resources (HR), it could mean coaching or laying off a worker.
- Preventive action processes begin with audits, investigations, and analyses of potential risks. Corrective action processes begin with identifying the causes of an occurring problem.
- While preventive actions are proactive, corrective actions are reactive.
- Preventive action processes begin with audits, investigations, and analyses of potential risks.

How is corrective action similar to preventive action?

Just as corrective action and preventive action are different, they also share certain similarities. Here are a few:

- Both are similar in intention to ensure the effective and efficient running of all functions of an organization.
- Both apply to all industries and any departments within them.
- Both fit nicely under the Plan–Do–Check–Act (PDCA) philosophy invented by the Deming–Shewhart cycle.
- Both sets of actions are Good Manufacturing Practices (GMPs).

Corrective action and preventive action in practice

We can now look at scenarios of corrective action and preventive action in the real world.

Let's say you're producing 5-m-long metal laminae and discover that some rods have different lengths. You first begin by investigating why this observation is so – the root cause. A bug in the software code is responsible for the defective manufacturing process.

You contact the machine supplier to request the correct code. The technicians do their job, and the machine now produces rods of the right length. It's a classic example of corrective action.

For preventive action, you want to begin producing rods. You may somehow be aware of an issue with the code for producing metal laminae and that this code would cause metal laminae to be different lengths.

Your production floor may prefer to ensure this defect does not occur, so they engage the managers to update the standard procedures to account for the different lengths. Even if they find that defects occur, the exercise itself is an illustration of preventive action to stop the defect from playing out in the future.

Preventive action shares similarities with risk management. Note that the latest version of the Quality Management System (QMS) standard, ISO 9001:2015, does not expect preventive action. Instead, the Quality Management System standard requires organizations to implement risk-based thinking (RBT).

Risk-based thinking requires you to pinpoint aspects with enough potential to impact the QMS where you are unsure of the outcome. It essentially involves:

- Identifying the risk or uncertainty.
- Determining if there's a need to take action to prevent undesirable outcomes or leverage positive results.

Implementing corrective and preventive action

Corrective or preventive actions should pass through change requests. The reason is that some of these actions may need a plan modification and some change in cost baseline. For the most part, there's not a significant impact on the cost baseline.

The organization bears the cost of corrective or preventive action. These are costs of quality, including the cost of conformance and the cost of nonconformance. The client doesn't pay for it.

While corrective action costs fall under the cost of nonconformance, preventive action costs are a cost of conformance.

The effectiveness of corrective and preventive action is crucially dependent on the systematic investigation of the root causes of failure. Good tools for root cause analysis include the five *Whys* and the *Ishikawa diagram*.

While corrective action costs fall under the cost of nonconformance, preventive action costs are a cost of conformance.

Using the corrective and preventive action subsystem

Here is a concise roundup of the purposes of the corrective and preventive action subsystem:

- Collection and analysis of information to identify existing and potential product and quality problems.
- Investigating product and quality problems and taking necessary and effective preventive and corrective action.

- Verification and validation of the effectiveness of corrective and preventive action.
- Communicating corrective and preventive action to the appropriate personnel.
- Providing information for management review.
- Documenting all activities.

BIBLIOGRAPHY

1. Sayana, S.A., 2003. Approach to auditing network security. *Information Systems Control Journal*, 5, pp.21–23.
2. Jackson, C., 2010. *Network security auditing*. Cisco Press.
3. Loughran, M., 2010. *Auditing for dummies*. John Wiley & Sons.
4. Russell, J.P., 2007. *The internal auditing pocket guide: Preparing, performing, reporting, and follow-up*. Quality Press.
5. Eilifsen, A., Messier, W.F., Glover, S.M. and Prawitt, D.F., 2014. *Auditing and assurance services*.
6. Leung, P., Coram, P. and Cooper, B., 2007. *Modern auditing & assurance services*. John Wiley & Sons.
7. Popa, M., 2010. Requirements for development of an assessment system for IT&C security audit. *Journal of Mobile, Embedded and Distributed Systems*, 2(2), pp.56–64.
8. Flowerday, S., Blundell, A.W. and Von Solms, R., 2006. Continuous auditing technologies and models: A discussion. *Computers & Security*, 25(5), pp.325–331.
9. Docs.oracle.com., 2022. *Contents*. [online] Available at: <https://docs.oracle.com/cd/E19056-01/sec.tk42/819-1503-10/index.html> [Accessed 10 March 2022].
10. Nagy, A.L., 2003. Auditing cases: An interactive learning approach. *Issues in Accounting Education*, 18(2), p.221.
11. Oswal, C.V., 2013. *Simplified approach to advanced auditing and professional ethics*. Walter Kluwer Pvt. Ltd.
12. Raj, A., 2016. A review on corrective action and preventive action (CAPA). *African Journal of Pharmacy and Pharmacology*, 10(1), pp.1–6.

Chapter 4

ISO 27001

4.1 OVERVIEW OF AN INFORMATION SECURITY AND MANAGEMENT SYSTEM

Security professionals, tasked with protecting the information assets of an organization, typically think of their responsibilities in three realms: confidentiality, integrity, and availability (CIA). The adversaries/attackers, seeking to disrupt an organization's security, have three corresponding goals in mind: disclosure, alteration, and denial (DAD). These models are known as the CIA and DAD Figure triads and are used by many security professionals around the world. The CIA and DAD triads are classic models of information security principles. Cybersecurity professionals use a well-known model to describe the goals of information security. The CIA triad is shown in Figure 4.1, which includes the three main characteristics of information that cybersecurity programs seek to protect.

- Confidentiality measures seek to prevent unauthorized access to information or systems.
- Integrity measures seek to prevent unauthorized modification of information or systems.
- Availability measures seek to ensure that legitimate use of information and systems remains possible.

Attackers or Pentester, and therefore penetration testers, seek to undermine these goals and achieve three corresponding goals of their own. The attacker's goals are known as the DAD.

When we talk about cybersecurity professionals, it is necessary to keep in mind that they need to have the knowledge of concepts about security, technical, and tools that are used day by day to defense and attack. That professional needs to have the mindset of an Attacker or Pentester, advanced knowledge about many kinds of attacks, as well as SQL Injection, Cross-Site Scripting (XSS), Cross-Site Scripting (XSS) Stored, Man-In-The-Middle (MITM), Brute-Force, Remote Code Execution, File Include, Directory or

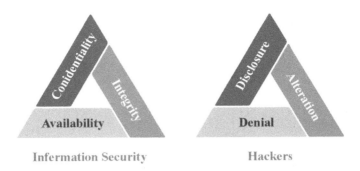

Figure 4.1 The CIA triad.

Path Traversal, Code Obfuscation, and more other concepts. The difference between those actors is the role that each one runs in an environment.

Information security is achieved by applying a suitable set of controls (policies, processes, procedures, organizational structures, and software and hardware functions). An information security management system (ISMS) is the way to protect and manage information based on a systematic business risk approach, to establish, implement, operate, monitor, review, maintain, and improve information security. It is an organizational approach to information security.

ISO publishes two standards that focus on an organization's ISMS:

- The code of practice standard: ISO 27002. This standard can be used as a starting point for developing an ISMS. It provides guidance for planning and implementing a program to protect information assets. It also provides a list of controls (safeguards) that you can consider implementing as part of your ISMS.
- The management system standard: ISO 27001. This standard is the specification for an ISMS. It explains how to apply ISO/IEC 27002. It provides the standard against which certification is performed, including a list of required documents. An organization that seeks certification of its ISMS is examined against this standard.

The standards set forth the following practices:

- All activities must follow a method. The method is arbitrary but must be well defined and documented.
- A company or organization must document its own security goals. An auditor will verify whether these requirements are fulfilled.

- All security measures used in the ISMS shall be implemented as the result of risk analysis in order to eliminate or reduce risks to an acceptable level.
- The standard offers a set of security controls. It is up to the organization to choose which controls to implement based on the specific needs of their business.
- A process must ensure the continuous verification of all elements of the security system through audits and reviews.
- A process must ensure the continuous improvement of all elements of the information and security management system.

The ISO 27001 standard adopts the Plan-Do-Check-Act (PDCA) model as its basis and expects the model will be followed in an ISMS implementation. These practices form the framework within which you will establish an ISMS. Refer Table 4.1, for further details.

Table 4.1 A sample visionary document to understand security policy for the information management and security and the objectives of the company in ensuring the security by undersigning

Security Policy
Protection of company assets is vital to the success of our business. To this end, we have established an information security management system that operates all the processes required to identify the information we need to protect and how we must protect it.
Because the needs of our business change, we recognize that our management system must be continually changed and improved to meet our needs. To this effect, we are continually setting new objectives and regularly reviewing our processes.
Objectives
It is the policy of our company to ensure:
- Information is only accessible to authorized persons from within or outside the company.
- Confidentiality of information is maintained.
- Integrity of information is maintained throughout the process.
- Business continuity plans are established, maintained, and tested.
- All personnel are trained on information security and are informed that compliance with the policy is mandatory.
- All breaches of information security and suspected weaknesses are reported and investigated.
- Procedures exist to support the policy, including virus control measures passwords, and continuity plans.
- Business requirements for the availability of information and systems will be met.
- The Information Security Manager is responsible for maintaining the policy and providing support and advice during its implementation.
- All managers are directly responsible for implementing the policy and ensuring staff compliance in their respective departments.

This policy has been approved by the company management and shall be reviewed by the management review team annually:
Signature:--- **Date:**---------------------
Title:-----------------------------

4.2 PURCHASE A COPY OF THE ISO/IEC STANDARDS

Before establishing an ISMS and drafting various documents for your ISMS, you should purchase copies of the pertinent ISO/IEC standards, namely:

The code of practice standard: ISO 27002:

This standard can be used as a starting point for developing an ISMS. It provides guidance for planning and implementing a program to protect information assets. It also provides a list of controls (safeguards) that you can consider implementing as part of your ISMS as illustrated in Figure 4.2.

The management system standard: ISO/IEC 27001:

This standard is the specification for an ISMS. It explains how to apply ISO 27002. It provides the standard against which certification is performed, including a list of required documents. An organization that seeks certification of its ISMS is examined against this standard.

Obtain management support as described in ISO/IEC 27001, management plays an important role in the success of an ISMS.

What you need: Management responsibility section of ISO 27001: Management must make a commitment to the establishment, implementation, operation, monitoring, review, maintenance, and improvement of the ISMS. Commitment must include activities such as ensuring that the proper resources are available to work on the ISMS and that all employees affected by the ISMS have the proper training, awareness, and competency.

Results: Establishment of the following items demonstrates management commitment:

- An information security policy: This policy can be a standalone document or part of an overall security manual that is used by an organization.
- Information security objectives and plans: This information can be a standalone document or part of an overall security manual that is used by an organization.
- Roles and responsibilities for information security: A list of the roles related to information security should be documented either in the organization's job description documents or as part of the security manual or ISMS description documents.
- Announcement or communication to the organization about the importance of adhering to the information security policy.
- Sufficient resources to manage, develop, maintain, and implement the ISMS.

In addition, management will participate in the ISMS Plan-Do-Check-Act [PDCA] process, as described in ISO 27001 by:

- Determining the acceptable level of risk. Evidence of this activity can be incorporated into the risk assessment documents, which are described later in this guide.

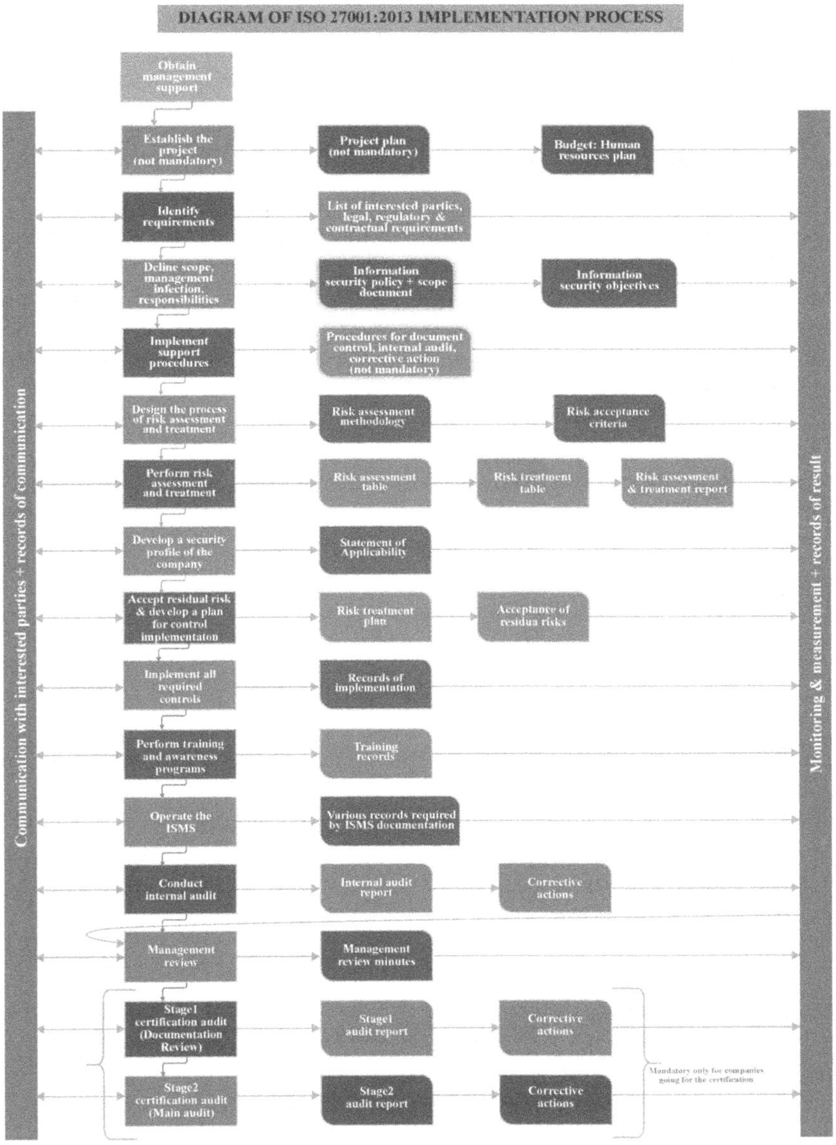

DIAGRAM OF ISO 27001:2013 IMPLEMENTATION PROCESS

Figure 4.2 Diagram of ISO 27001:2013 implementation process.

- Conducting management reviews of the ISMS at planned intervals. Evidence of this activity can be part of the approval process for the documents in the ISMS.
- Ensuring that personnel affected by the ISMS are provided with training, are competent for the roles and responsibilities they are assigned

to fulfill and are aware of those roles and responsibilities. Evidence of this activity can be through employee training records and employee review documents.

4.3 DETERMINE THE SCOPE OF THE ISMS

When management has made the appropriate commitments, you can begin to establish your ISMS. In this step, you should determine the extent to which you want the ISMS to apply to your organization. The scope and purpose are mentioned in Table 4.2.

What you need: You can use several of the "result" documents that were created as part of step two, such as:

- The information security policy.
- The information security objectives and plans.
- The roles and responsibilities that are related to information security and were defined by the management.

In addition, you will need:

- Lists of the areas, locations, assets, and technologies of the organization that will be controlled by the ISMS.
- What areas of your organization will be covered by the ISMS?
- What are the characteristics of those areas, its locations, assets, and technologies to be included in the ISMS?
- Will you require your suppliers to abide by your ISMS?
- Are there dependencies on other organizations? Should they be considered?

Table 4.2 A sample copy of the information security management system defining the scope and purpose of ISO/IEC 27001:2005

Scope and Purpose
The company is committed to protecting its information and that of its customers
To achieve this goal, the company has implemented an Information Security Management System in accordance with ISO/IEC 27001:2005
The company's ISMS is applicable to the following areas of the business:
  • Finance department
  • Internal IT systems and networks used for back-end business (such as e-mail, timesheets, contract development and storage, report writing)
"Note: IT systems on which company software is developed and stored are part of the Software Development ISMS. Refer to the Software Development Security Manual for more information

Your goals will be to cover the following:

- The processes used to establish the scope and context of the ISMS.
- The strategic and organizational context.

Important: Keep your scope manageable. Consider including only parts of the organization, such as a logical or physical grouping within the organization. Large organizations might need several Information Security Management Systems in order to maintain manageability. For example, they might have one ISMS for their finance department and the networks used by that department and a separate ISMS for their software development department and systems.

Results: A documented scope for your ISMS.

When you have determined the scope, you will need to document it, usually in a few statements or paragraphs. The documented scope often becomes one of the first sections of your organization's security manual. Or, it might remain a standalone document in a set of ISMS documents that you plan to maintain. Often the scope, the security policy, and the security objectives are combined into one document.

4.4 IDENTIFY APPLICABLE LEGISLATION

After you have determined the scope, identify any regulatory or legislative standards that apply to the areas you plan to cover with the ISMS. Such standards might come from the industry in which your organization works or from the state, local, or federal governments, or international regulatory bodies.

What you need: Up-to-date regulatory or legislative standards that might be applicable to your organization. You might find it helpful to have input and review from lawyers or specialists who are knowledgeable about the standards.

Results: Additional statements in the scope of the ISMS. If your ISMS will incorporate more than two or three legislative or regulatory standards, you might also create a separate document or appendix in the security manual that lists all of the applicable standards and details about the standards.

Example: The text added to the scope statement as a result of identifying applicable legislation is shown in Table 4.2 and explained with the following example.

Scope and purpose

The company is committed to protecting its information and that of its customers. To achieve this goal, the company has implemented an information security management system in accordance with ISO 27001: 2013 and the rules and regulations that are part of the Information Technology Act, 2000 also known as IT Act.

The company's ISMS is applicable to the following areas of the business:

- Finance department.
- Internal IT systems and networks used for back-end business (such as e-mail, timesheets, contract development and storage, and report writing).

4.5 DEFINE A METHOD OF RISK ASSESSMENT

Risk assessment is the process of identifying risks by analyzing threats to, impacts on, and vulnerabilities of information and information systems and processing facilities, and the likelihood of their occurrence. Choosing a risk assessment method is one of the most important parts of establishing an ISMS. To meet the requirements of ISO 27001, you will need to define and document a method of risk assessment and then use it to assess the risk to your identified information assets, make decisions about which risks are intolerable and therefore need to be mitigated, and manage the residual risks through carefully considered policies, procedures, and controls.

ISO does not specify the risk assessment method you should use; however, it does state that you must use a method that enables you to complete the following tasks:

- Evaluate risk based on levels of confidentiality, integrity, and availability. Some risk assessment methods provide a matrix that defines levels of confidentiality, integrity, and availability and provides guidance as to when and how those levels should be applied, as shown in the following table:
- Set objectives to reduce risk to an acceptable level.
- Determine criteria for accepting the risk.
- Evaluate risk treatment options.

There are many risk assessment methods you can choose from, such as those that are prevalent in your industry. For example, if your company is in the oil industry, you might find there are risk assessment methods related to that industry. Table 4.3 shows details about the risk and the impact factor of risk.

When you have completed this step, you should have a document that explains how your organization will assess risk, including:

- The organization's approach to information security risk management;
- Criteria for information security risk evaluation and the degree of assurance required.

Table 4.3 Impact of loss on integrity, confidentiality, and availability as low, medium, and high

Impact of loss	Low	Medium	High
Confidentiality Ensuring that information is accessible only to those authorized to have access	The unauthorized disclosure of information could be expected to have a limited adverse effect on organizational operations, organizational assets, or individuals.	The unauthorized disclosure of information could be expected to have a serious adverse effect on organizational operations, organizational assets, or individuals.	The unauthorized disclosure of information could be expected to have a severe or catastrophic adverse effect on organizational operations, organizational assets, or individuals.
Integrity Safeguarding the accuracy and completeness of information and processing methods	The unauthorized modification or destruction of information could be expected to have a limited adverse effect on organizational operations, organizational assets, or individuals.	The unauthorized modification or destruction of information could be expected to have a serious adverse effect on organizational operations, organizational assets, or individuals.	The unauthorized modification or destruction of information could be expected to have a severe or catastrophic adverse effect on organizational operations, organizational assets, or individuals.
Availability Ensuring that authorized users have access to information and associated assets when required	The disruption of access to or use of information or an information system could be expected to have a limited adverse effect on organizational operations, organizational assets, or individuals.	The disruption of access to or use of information or an information system could be expected to have a serious adverse effect on organizational operations, organizational assets, or individuals.	The disruption of access to or use of information or an information system could be expected to have a severe or catastrophic adverse effect on organizational operations, organizational assets, or individuals.

4.6 CREATE AN INVENTORY OF INFORMATION ASSETS TO PROTECT

To identify risks and the levels of risks associated with the information you want to protect, you first need to make a list of all of your information assets that are covered in the scope of the ISMS.

What you will need: You will need the scope that you defined in Step 3 and input from the organization that is defined in your scope regarding its information assets.

Table 4.4 Risk assessment for asset table with placeholder

Assert	Details	Owner	Location	CIA profile	Replacement value
Strategic information	Medium- and long-term plans	CEO	CEO PC		High
Project plans	Short-term plans	CEO	CEO PC		Medium

Result: When you have completed this step, you should have a list of the information assets to be protected and an owner for each of those assets. You might also want to identify where the information is located and how critical or difficult it would be to replace. This list should be part of the risk assessment methodology document that you created in the previous step.

Because you will need this list to document your risk assessment, you might want to group the assets into categories and then make a table of all the assets with columns for assessment information and the controls you choose to apply. The following examples show an asset Table 4.4.

4.7 IDENTIFY RISKS

Next, for each asset you defined in the previous step, you will need to identify risks and classify them according to their severity and vulnerability. In addition, you will need to identify the impact that loss of confidentiality, integrity, and availability may have on the assets.

To begin identifying risks, you should start by identifying actual or potential threats and vulnerabilities for each asset. A threat is something that could cause harm. For example, a threat could be any of the following:

- A declaration of the intent to inflict harm or misery.
- Potential to cause an unwanted incident, which may result in harm to a system or organization and its assets.
- The intentional, accidental, or man-made act that could inflict harm or an act of God (such as a hurricane or tsunami).

A vulnerability is a source or situation with a potential for harm (for example, a broken window is a vulnerability; it might encourage harm, such as a break-in). A risk is a combination of the likelihood and severity or frequency that a specific threat will occur.

What you will need:

- The list of assets.
- The risk assessment methodology.

Table 4.5 Risk assessment for asset table with placeholder

Assert	Details	Owner	Location	CIA profile	Replacement value	Risk summary
Strategic information	Medium and long-term plans	CEO	CEO PC	C-High I: High A: Medium	High	Disclosure (gives advantage to third party)
Project plans	Short-term plans	CEO	CEO PC	C: High I: High A: Low	Medium	Disclosure (gives advantage to competitor): company might lose business

For each asset, you should identify vulnerabilities that might exist for that asset and threats that could result from those vulnerabilities. It is often helpful to think about threats and vulnerabilities in pairs, with at least one pair for each asset and possibly multiple pairs for each asset.

Results: For each asset, you will have a threat and vulnerability description, and using your risk assessment methodology, you will assign levels of confidentiality, integrity, and availability to that asset. If you used a table, you can add this information to that table, as shown in the following example.

Example: The risk summary column describes the threat and vulnerability. The CIA profile classifies the asset's confidentiality, integrity, and availability. Refer to Table 4.5.

4.8 ASSESS THE RISKS

After you have identified the risks and the levels of confidentiality, integrity, and availability, you will need to assign values to the risks. The values will help you determine if the risk is tolerable or not and whether you need to implement a control to either eliminate or reduce the risk. To assign values to risks, you need to consider:

- The value of the asset being protected.
- The frequency with which the threat or vulnerability might occur.
- The damage that the risk might inflict on the company or its customers or partners.

For example, you might assign values of low, medium, and high to your risks. To determine which value to assign, you might decide that if the value of an asset is high and the damage from a specified risk is high, the value of the risk should also be high, even though the potential frequency is low. Your risk assessment methodology document should tell you what values to use and might also specify the circumstances under which specific values should be assigned. Also, be sure to refer to your risk assessment methodology document to determine the implication of a certain risk value. For example, to keep your ISMS manageable, your risk assessment methodology might specify that only risks with a value of medium or high will require control in your ISMS. Based on your business needs and industry standards, risk will be assigned appropriate values.

What you will need:

- Lists of assets and their associated risks and CIA levels, which you created in the previous step.
- Possibly input from management as to what level of risk they are willing to accept for specific assets.

Results: When you have completed your assessment, you will have identified which information assets have intolerable risk and therefore require controls. You should have a document (sometimes referred to as a risk assessment report) that indicates the risk value for each asset. In the next step, you will identify which controls might be applicable for the assets that require control in order to reduce the risk to tolerable levels. This document can either be standalone or it can be part of an overall risk assessment document that contains your risk assessment methodology and this risk assessment.

Examples: If you used a table similar to the one in the preceding examples, your result after completing this step might look like the following Table 4.6 example.

4.9 IDENTIFY APPLICABLE OBJECTIVES AND CONTROLS

Next, for the risks that you've determined to be intolerable, you must take one of the following actions:

- Decide to accept the risk, for example, actions are not possible because they are out of your control (such as natural disaster or political uprising) or are too expensive.
- Transfer the risk, for example, purchase insurance against the risk, subcontract the activity so that the risk is passed on to the subcontractor, etc.
- Reduce the risk to an acceptable level through the use of controls.

Table 4.6 Risk assessment for asset table with placeholder

Asset	Details	Owner	Location	CIA Profile	Replacement Value	Risk Summary	Risk Value
Strategic information	Medium and long-term plans	CEO	CEO PC	C: High I: High A: Medium	High	Disclosure (gives advantage to third party)	High
Project plans	Short-term plans	CEO	CEO PC	C: High I: High A: Low	Medium	Disclosure (gives advantage to competitor); company might lose business	Medium
HR documents	Employee records	Company board	HR management company	C: High I: High A: Low	Medium	Disclosure of personal information	Medium

To reduce the risk, you should evaluate and identify appropriate controls. These controls might be controls that your organization already has in place or controls that are defined in the ISO 27002 standard.

What you will need:

- Annex A of ISO 27001. This appendix summarizes controls that you might want to choose from.
- ISO 27002, which provides greater detail about the controls summarized in ISO 27001.
- Procedures for existing corporate controls.

Results: You should end up with two documents by completing this step:

- A Risk Treatment Plan.
- A Statement of Applicability.

The Risk Treatment Plan documents the following:

- The method selected for treating each risk (accept, transfer, reduce).
- Which controls are already in place?
- What additional controls are proposed?
- The time frame over which the proposed controls are to be implemented.

The Statement of Applicability (SOA) documents the control objectives and controls selected from Annex A. The Statement of Applicability is usually a large table in which each control from Annex A of ISO/IEC 27001 is listed with its description and corresponding columns that indicate whether that control was adopted by the organization, the justification for adopting or not adopting the control, and a reference to the location where the organization's procedure for using that control is documented.

The SOA can be part of the risk assessment document, but usually, it is a standalone document because it is lengthy and is listed as a required document in the standard. For additional help with creating a Risk Treatment Plan and a Statement of Applicability, refer to the example in Table 4.7.

Examples of Risk Treatment Plan:

If you used a table as described in the preceding steps, the control analysis portion of your Risk Treatment Plan could be covered by the Control column and the Sufficient Control column, as shown in the following example. Any risks that you transfer to others or that you choose to accept as they are should also be recorded in your treatment plan.

The remaining Risk Treatment Plan requirements could be met by adding this table and by explaining the methods used for treating risk and the time frame in which the controls will be implemented to a risk assessment methodology document, like the one you created in step 4.

Table 4.7 Risk assessment for asset table with placeholder

Assert	Details	Owner	Location	CIA profile	Replacement value	Risk summary	Risk value	Control
Strategic information	Medium and long-term plans	CEO	CEO PC	C: High I: High A: Medium	High	Disclosure (gives advantage to third party)	High	15.1.1
Project plans	Short-term plans	CEO	CEO PC	C: High I: High A: Low	Medium	Disclosure (gives advantage to competitor); company might lose business	Medium	15.1.1
HR documents	Employee records	Company board	HR management company	C: High I: High A: Low	Medium	Disclosure of personal information	Medium	None: HR activities and documentation management outsourced

4.10 SET UP POLICY, PROCEDURES, AND DOCUMENTED INFORMATION TO CONTROL RISKS

For each control that you define, you must have corresponding statements of policy or in some cases a detailed procedure. The procedure and policies are used by affected personnel so they understand their roles and so that the control can be implemented consistently. The documentation of the policy and procedures is a requirement of ISO 27001.

What you will need: To help you identify which procedures you might need to document, refer to your Statement of Applicability. To help you write your procedures so that they are consistent in content and appearance, you might want to create some type of template for your procedure writers to use.

Results: Additional policy and documented information. (The number of documents you produce will depend on the requirements of your organization.) Some of these procedures might also generate records. For example, if you have a procedure that all visitors to your facility must sign a visitor's log, the log itself becomes a record providing evidence that the procedure has been followed.

Example: The number of policies, procedures, and records that you will require as part of your ISMS will depend on a number of factors, including the number of assets you need to protect and the complexity of the controls you need to implement. The example that follows shows a partial list of one organization's set of documents:

Mandatory documents and records required by ISO 27001:2013

- Scope of the ISMS (clause 4.3)
- Information security policy and objectives (clauses 5.2 and 6.2)
- Risk assessment and risk treatment methodology (clause 6.1.2)
- Statement of Applicability (clause 6.1.3 d)
- Risk Treatment Plan (clauses 6.1.3 e and 6.2)
- Risk assessment report (clause 8.2)
- Definition of security roles and responsibilities (clauses A.7.1.2 and A.13.2.4)
- Inventory of assets (clause A.8.1.1)
- Acceptable use of assets (clause A.8.1.3)
- Access control policy (clause A.9.1.1)
- Operating procedures for IT management (clause A.12.1.1)
- Secure system engineering principles (clause A.14.2.5)
- Supplier security policy (clause A.15.1.1)
- Incident management procedure (clause A.16.1.5)
- Business continuity procedures (clause A.17.1.2)
- Statutory, regulatory, and contractual requirements (clause A.18.1.1)

And here are the mandatory records:

- Records of training, skills, experience, and qualifications (clause 7.2)
- Monitoring and measurement results (clause 9.1)
- Internal audit program (clause 9.2)
- Results of internal audits (clause 9.2)
- Results of the management review (clause 9.3)
- Results of corrective actions (clause 10.1)
- Logs of user activities, exceptions, and security events (clauses A.12.4.1 and A.12.4.3)

Non-mandatory documents

There are numerous non-mandatory documents that can be used for ISO 27001 implementation, especially for the security controls from Annex A. However, I find these non-mandatory documents to be most commonly used:

- Procedure for document control (clause 7.5)
- Controls for managing records (clause 7.5)
- Procedure for internal audit (clause 9.2)
- Procedure for corrective action (clause 10.1)
- Bring your own device (BYOD) policy (clause A.6.2.1)
- Mobile device and teleworking policy (clause A.6.2.1)
- Information classification policy (clauses A.8.2.1, A.8.2.2, and A.8.2.3)
- Password policy (clauses A.9.2.1, A.9.2.2, A.9.2.4, A.9.3.1, and A.9.4.3)
- Disposal and destruction policy (clauses A.8.3.2 and A.11.2.7)
- Procedures for working in secure areas (clause A.11.1.5)
- Clear desk and clear screen policy (clause A.11.2.9)
- Change management policy (clauses A.12.1.2 and A.14.2.4)
- Backup policy (clause A.12.3.1)
- Information transfer policy (clauses A.13.2.1, A.13.2.2, and A.13.2.3)
- Business impact analysis (clause A.17.1.1)
- Exercising and testing plan (clause A.17.1.3)
- Maintenance and review plan (clause A.17.1.3)
- Business continuity strategy (clause A.17.2.1)

4.11 ALLOCATE RESOURCES AND TRAIN THE STAFF

Adequate resources (people, time, money) should be allocated to the operation of the ISMS and all security controls. In addition, the staff who must work within the ISMS (maintaining it and its documentation and implementing its controls) must receive appropriate training. The success of the training program should be monitored to ensure that it is effective.

Therefore, in addition to the training program, you should also establish a plan for how you will determine the effectiveness of the training.

What you will need:

- A list of the employees who will work within the ISMS.
- All of the ISMS procedures to use for identifying what type of training is needed and which members of the staff or interested parties will require training.
- Management agreement to the resource allocation and the training plans.

Results:

Specific documentation is not required in the ISO/IEC standards. However, to provide evidence that resource planning and training has taken place, you should have some documentation that shows who has received training and what training they have received. In addition, you might want to include a section for each employee that lists what training they should be given. Also, you will probably have some type of procedure for determining how many people, how much money, and how much time needs to be allocated to the implementation and maintenance of your ISMS. It's possible that this procedure already exists as part of your business operating procedures or that you will want to add an ISMS section to that existing documentation.

4.12 MONITOR THE IMPLEMENTATION OF THE ISMS

To ensure that the ISMS is effective and remains current, suitable, adequate, and effective, ISO 27001 requires:

- Management to review the ISMS at planned intervals. The review must include assessing opportunities for improvement, and the need for changes to the ISMS, including the security policy and security objectives, with specific attention to previous corrective or preventative actions and their effectiveness.
- Periodic internal audits. The results of the reviews and audits must be documented and records related to the reviews and audits must be maintained.

What you will need:

To perform management reviews, ISO 27001 requires the following input:

- Results of ISMS internal and external audits and reviews.
- Feedback from interested parties.

- Techniques, products, or procedures which could be used in the organization to improve the effectiveness of the ISMS.
- Preventative and corrective actions (including those that might have been identified in previous reviews or audits).
- Incident reports, for example, if there has been a security failure, a report that identifies what the failure was, when it occurred, and how it was handled and possibly corrected.
- Vulnerabilities or threats not adequately addressed in the previous risk assessment.
- Follow-up actions from previous reviews.
- Any organizational changes that could affect the ISMS.
- Recommendations for improvement.

To perform internal audits on a periodic basis, you need to define the scope, criteria, frequency, and methods. You also need the procedure (which should have been written as part of step 10) that identifies the responsibilities and requirements for planning and conducting the audits and for reporting results and maintaining records.

Results:
The results of a management review should include decisions and actions related to:

- Improvements to the ISMS.
- Modification of procedures that affect information security at all levels within the organization.
- Resource needs.
- The results of an internal audit should result in the identification of nonconformities and their related corrective actions or preventative actions. ISO 27001 lists the activity and record requirements related to corrective and preventative actions.

4.13 PREPARE FOR THE CERTIFICATION AUDIT

If you plan to have your ISMS certified, you will need to conduct a full cycle of internal audits, management review, and activities in the PDCA process. The external auditor will first examine your ISMS documents to determine the scope and content of your ISMS. Then the auditor will examine the necessary records and evidence that you implement and practice what is stated in your ISMS.

What you will need:

- All of the documents that you created in the preceding steps.
- Records from at least one full cycle of management reviews, internal audits, and PDCA activities, and evidence of responses taken as the result of those reviews and audits.

Results:

The results of this preparation should be a set of documents that you can send to an auditor for review and a set of records and evidence that will demonstrate how efficiently and completely you have implemented your ISMS.

BIBLIOGRAPHY

1. Hamdi, Z., Norman, A.A., Molok, N.N.A. and Hassandoust, F., 2019, December. A comparative review of {ISMS} implementation based on {ISO} 27000 series in organizations of different business sectors. *Journal of Physics: Conference Series, 1339*(1), p.12103.
2. Hardy, C.A. and Williams, S.P., 2010. Managing information risks and protecting information assets in a web 2.0 era. In Bled eConference (p.25).
3. Gillies, A., 2011. Improving the quality of information security management systems with ISO27000. *TQM Journal*.
4. Calkins, H. et al., 2007. HRS/EHRA/ECAS expert consensus statement on catheter and surgical ablation of atrial fibrillation: Recommendations for personnel, policy, procedures and follow-up: A report of the Heart Rhythm Society (HRS) task force on catheter and surgical ablation of atrial fibrillation developed in partnership with the European Heart Rhythm Association (EHRA) and the European Cardiac Arrhythmia Society (ECAS); in collaboration with the American College of Cardiology (ACC), American Heart Association (AHA), and the Society of Thoracic Surgeons (STS). Endorsed and approved by the governing bodies of the American College of Cardiology, the American Heart Association, the European Cardiac Arrhythmia Society, the European Heart Rhythm Association, the Society of Thoracic Surgeons, and the Heart Rhythm Society. *Europace, 9*(6), pp.335–379.
5. Peltier, T.R., 2005. *Information security risk analysis*. CRC press.
6. Al-Mayahi, I. and Sa'ad, P.M., 2012. Iso 27001 gap analysis-case study. In Proceedings of the International Conference on Security and Management (SAM) (p. 1). The Steering Committee of The World Congress in Computer Science, Computer Engineering and Applied Computing (WorldComp).
7. Docs.oracle.com. 2022. *Contents*. [online] Available at: <https://docs.oracle.com/cd/E19056-01/sec.tk42/819-1503-10/index.html> [Accessed 10 March 2022].
8. Calder, A., 2017. *Nine steps to success: An ISO 27001 implementation overview*. IT Governance Ltd.
9. Carlson, T. and Forbes, R., 2010. A business case for ISO 27001 certification. In *Information security management handbook*, Volume 4 (pp. 153–160). Auerbach Publications.
10. Sharma, N.K. and Dash, P.K., 2012. Effectiveness of ISO 27001, as an information security management system: An analytical study of financial aspects. *Far East Journal of Psychology and Business, 9*(3), pp.42–55.
11. Kenyon, B., 2019. *ISO 27001 controls: A guide to implementing and auditing*. IT Governance Ltd.
12. Χατζηδημητρίου, E., 2019. The implementation of information security policy using ISO 27001: A case study in a software company.

… Chapter 5

Asset Management

5.1. WHAT ARE ASSETS ACCORDING TO ISO 27001?

Since ISO 27001 focuses on the preservation of confidentiality, integrity, and availability of information, this means that assets can be:

1. Hardware: e.g. not only laptops, servers, printers but also mobile phones or USB memory sticks.
2. Software: not only the purchased software but also freeware.
3. Information: not only in electronic media (databases, files in PDF, Word, Excel, and other formats) but also in paper and other forms.
4. Infrastructure: e.g. offices, electricity, air conditioning – because those assets can cause a lack of availability of information. People are also considered assets because they also have lots of information in their heads, which is very often not available in other forms.
5. Outsourced services: e.g. not only legal services or cleaning services but also online services like Dropbox or Gmail – it is true that these are not assets in the pure sense of the word, but such services need to be controlled very similarly to assets, so they are very often included in the asset management.

5.2. WHY ARE ASSETS IMPORTANT FOR INFORMATION SECURITY MANAGEMENT?

There are two reasons why managing assets is important:

1. Assets are usually used to perform the risk assessment: Although not mandatory by ISO 27001:2013, assets are usually the key element of identifying risks, together with threats and vulnerabilities.
2. If the organization doesn't know who is responsible for which asset, chaos would ensue: Defining asset owners and assigning them the responsibility to protect the confidentiality, integrity, and availability of the information is one of the fundamental concepts in ISO 27001.

This is why ISO 27001:2013 requires the following: An inventory of assets needs to be developed (A.8.1.1), owners of the assets need to be nominated (A.8.1.2), and acceptable use of assets must be defined (A.8.1.3).

5.3. HOW TO BUILD AN ASSET INVENTORY?

If you didn't develop your asset inventory previously, the easiest way to build is during the initial risk assessment process (if you have chosen the asset-based risk assessment methodology) because this is when all the assets need to be identified, together with their owners.

The best way to build asset inventory is to interview the head of each department and list all the assets a department uses. The easiest way is "describe-what-you-see" technique; basically, ask the head of each department to list all the software that he or she sees that are installed on the computer, all the documents in their folders and file cabinets, all the people working in the department, all the equipment seen in their offices, etc.

Of course, if you already do have some existing asset inventories (e.g. fixed asset register, employee list, licensed software list, etc.), then you don't have to duplicate those lists, the best would be to refer to your other lists from your information security asset register.

ISO 27001 does not prescribe which details must be listed in the asset inventory, you can list only the asset name and its owner, but you can also add some other useful information, like asset category, its location, and some notes.

Building the asset register is usually done by the person who coordinates the ISO 27001 implementation project – in most cases, this is the Chief Information Security Officer, and this person collects all the information and makes sure that the inventory is updated.

5.4. WHO SHOULD BE THE ASSET OWNER?

The owner is normally a person who operates the asset and who makes sure the information related to this asset is protected. For instance, an owner of a server can be the system administrator, and the owner of a file can be the person who has created this file; for the employees, the owner is usually the person who is their direct supervisor.

For similar assets used by many people (such as laptops or mobile phones), you can define that an asset owner is the person using the asset, and if you have a single asset used by many people (e.g. an ERP software), then an asset owner can be a member of the board who has the responsibility throughout the whole organization – in this case of ERP, this could be the Chief Information Officer.

5.5. ISO 27001/ISO 27005 RISK ASSESSMENT & TREATMENT – SIX BASIC STEPS

Risk assessment (often called risk analysis) is probably the most complex part of ISO 27001 implementation, but at the same time, risk assessment (and treatment) is the most important step at the beginning of your information security project – it sets the foundations for information security in any company.

Why is it so important? The main philosophy of ISO 27001 is to find out which incidents could occur (i.e. assess the risks) and then find the most appropriate ways to avoid such incidents (i.e. treat the risks). Not only this, but one also has to assess the importance of each risk so that one can focus on the most important ones. Although risk assessment and treatment (together: risk management) is a complex job

5.6. THE BASIC STEPS WILL SHED LIGHT ON WHAT ONE HAS TO DO

5.6.1 ISO 27001 risk assessment methodology

This is the first step on your voyage through risk management. You need to define rules on how you are going to perform the risk management because you want your whole organization to do it the same way – the biggest problem with risk assessment happens if different parts of the organization perform it in a different way. Therefore, you need to define whether you want qualitative or quantitative risk assessment, which scales you will use for qualitative assessment, what will be the acceptable level of risk, etc.

5.6.2 Risk assessment implementation

Once you know the rules, you can start finding out which potential problems could happen to you – you need to list all your assets, then threats and vulnerabilities related to those assets, and assess the impact and likelihood for each combination of assets/threats/vulnerabilities, and finally calculate the level of risk.

Companies are usually aware of only 30% of their risks. Therefore, you'll probably find this kind of exercise quite revealing – when you are finished you'll start to appreciate the effort you've made.

5.6.3 Risk treatment implementation

Of course, not all risks are created equal – you have to focus on the most important ones, so-called unacceptable risks.

There are four options you can choose from to mitigate each unacceptable risk:

1. Apply security controls from Annex A to decrease the risks – we discuss **ISO 27001 Annex A controls** in the next section.
2. Transfer the risk to another party – e.g. to an insurance company by buying an insurance policy.
3. Avoid the risk by stopping an activity that is too risky or by doing it in a completely different fashion.
4. Accept the risk – if the cost for mitigating that risk would be higher than the damage itself.

5.6.4 ISMS risk assessment report

Unlike previous steps, this one is quite boring – you need to document everything you've done so far. Not only for the auditors but also you may want to check yourself these results in a year or two.

5.6.5 Statement of applicability

This document actually shows the security profile of your company – based on the results of the risk treatment, you need to list all the controls you have implemented, why you have implemented them, and how. This document is also very important because the certification auditor will use it as the main guideline for the audit.

5.6.6 Risk treatment plan

This is the step where you have to move from theory to practice.

The purpose of the risk treatment plan is to define exactly who is going to implement each control, in which timeframe, with which budget, etc. I would prefer to call this document "implementation plan" or "action plan".

Once you've written this document, it is crucial to get your management approval because it will take considerable time and effort (and money) to implement all the controls that you have planned here. And without their commitment you won't get any of these.

5.7. ISO 27001 CONTROLS FROM ANNEX A

Annex A of **ISO 27001** is probably the most famous annex of all the ISO standards – this is because it provides an essential tool for managing information security risks: a list of security controls (or safeguards) that are to be used to improve the security of information assets.

This section will provide an understanding of how Annex A is structured, as well as its relationship with the main part of ISO 27001 and with ISO 27002.

5.7.1 How many domains are there in ISO 27001?

The ISO 27001 controls list can be found in Annex A, and it is organized into 14 sections (domains). Below you can find a breakdown of what particular sections are focused on:

1. Sections related to organizational issues: A.5, A.6., A.8, A.15
2. Section related to human resources: A.7
3. IT-related sections: A.9, A.10, A.12, A.13. A.14, A.16, A.17
4. Section related to physical security: A.11
5. Section related to legal issues: A.18

5.7.2 What are the 14 domains of ISO 27001?

Here's a short description of each of the 14 sections:
Figure 5.1 shows the various domains of ISO 27001.

- A.5 Information security policies: Controls on how the policies are written and reviewed.
- A.6 Organization of information security: Controls on how the responsibilities are assigned; also includes the controls for mobile devices and teleworking.
- A.7 Human resources security: Controls prior to employment, during, and after the employment.
- A.8 Asset management: Controls related to inventory of assets and acceptable use; also, for information classification and media handling.
- A.9 Access control: Controls for the management of access rights of users, systems, and applications and for the management of user responsibilities.
- A.10 Cryptography: Controls related to encryption and key management.
- A.11 Physical and environmental security: Controls defining secure areas, entry controls, protection against threats, equipment security, secure disposal, Clear Desk and Clear Screen Policy, etc.
- A.12 Operational security: Lots of controls related to the management of IT production: change management, capacity management, malware, back-up, logging, monitoring, installation, vulnerabilities, etc.
- A.13 Communications security: Controls related to network security, segregation, network services, transfer of information, messaging, etc.
- A.14 System acquisition, development, and maintenance: Controls defining security requirements and security in development and support processes.

Figure 5.1 Various domains of ISO 27001.

- A.15 Supplier relationships: Controls on what to include in agreements and how to monitor the suppliers.
- A.16 Information security incident management: Controls for reporting events and weaknesses, defining responsibilities, response procedures, and collection of evidence.
- A.17 Information security aspects of business continuity management: Controls requiring the planning of business continuity, procedures, verification and reviewing, and IT redundancy.
- A.18 Compliance: Controls requiring the identification of applicable laws and regulations, intellectual property protection, personal data protection, and reviews of information security.

5.8. THE IMPORTANCE OF STATEMENT OF APPLICABILITY FOR ISO 27001

The importance of the Statement of Applicability in ISO 27001 (sometimes referred to as SoA) is usually underrated – like the Quality Manual in ISO

9001, it is the central document that defines how you will implement a large part of your information security.

Actually, the Statement of Applicability (ISO 27001 Clause 6.1.3 d) is the main link between the risk assessment and treatment and the implementation of your information security – its purpose is to define which of the suggested 114 controls (security measures) from **ISO 27001** Annex A you will apply, and for those that are applicable will then be implemented. As Annex A is considered to be comprehensive, but not exhaustive for all situations, nothing prevents you from also considering another source for the controls.

5.8.1 Why it is needed?

Now, why is such a document necessary when you already produced the Risk Assessment Report (which is also mandatory) which also defines the necessary controls? Here are the reasons:

- First of all, during risk treatment, you identify the controls that are necessary because you identified risks that need to be decreased; however, in SoA, you also identify the controls that are required due to other reasons – i.e. because of the law, contractual requirements, and other processes, etc.
- Second, the Statement of Applicability justifies the inclusion and exclusion of controls from Annex A and the inclusion of controls from another source.
- Third, the Risk Assessment Report could be quite lengthy – some organizations might identify a few thousand risks (sometimes even more), so such a document is not really useful for everyday operational use; on the other hand, the Statement of Applicability is rather short – it has a row for each control (114 from Annex A, plus the added ones), which makes it possible to present it to management and to keep it up-to-date.
- Fourth, and most important, SoA must document whether each applicable control is already implemented or not. Good practice (and most auditors will be looking for this) is also to describe how each applicable control is implemented – e.g. either by making a reference to a document (policy/procedure/working instruction, etc.) or by shortly describing the procedure in use or equipment that is used.

Actually, if you go for the ISO 27001 certification, the certification auditor will take your Statement of Applicability and walk around your company checking out whether you have implemented your controls in the way you described them in your SoA. It is the central document for doing their on-site audit.

A very small number of companies realize that by writing a good Statement of Applicability you could decrease the number of other documents – for instance, if you want to document a certain control, but if the description of the procedure for that control would be rather short, you can describe it in the SoA. Therefore, you would avoid writing another document.

5.9. ISO 27001: A.8 ASSET MANAGEMENT

5.9.1 Introduction

An asset is an item of value. An asset is defined as *"Any item of economic value owned by an individual or corporation"*. It can be referring to items such as buildings, utility infrastructure such as electrical cables, water pipes, rail lines, and metro tunnels, and industrial assets such as oil rigs, chemical plants, and process plant conveyors. Asset and data management is based on the idea that it is important to identify, track, classify, and assign ownership to the most important assets in the organization to ensure they are adequately protected. Tracking the inventory of IT hardware is the simplest example of asset management. Knowing what you have, where it lives, how important it is, and who's responsible for it. An information asset is an item of value containing information. The same concepts of general asset management apply to the management of information assets (e.g., data). To be effective, an overall asset management strategy should include information assets, software assets, and information technology equipment. In addition, the people employed by an organization, as well as the organization's reputation, are also important assets not to be overlooked in an effective asset management strategy.

The asset types and the vital understanding of their business context are illustrated in Figure 5.2. The organization should recognize

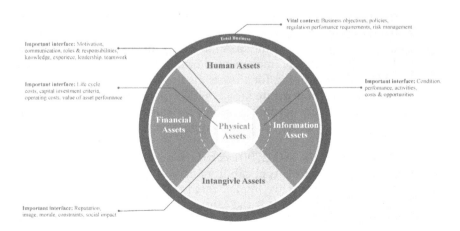

Figure 5.2 The asset types and the vital understanding.

such interdependencies and make appropriate provisions for the indirect "enablers" that are required to optimize the value of physical assets. Conversely, organizations that are heavily dependent upon physical assets should also recognize that deficiencies in the management of other asset types may have a profound impact on the overall or long-term performance of their physical assets and thus their organizational performance.

Such organizations should recognize that all assets will need to be managed in an integrated and holistic manner.

For example:

- Human assets: The behaviors, knowledge, and competence of the workforce have a fundamental influence on the performance of the physical assets.
- Financial assets: Financial resources are required for infrastructure investments, operation, maintenance, and materials.
- Information assets: Good quality data and information are essential to develop, optimize, and implement the asset management plan.
- Intangible assets: The organization's reputation and image can have a significant impact on infrastructure investment, operating strategies, and associated costs.

An organization should be in a position to know what physical, environmental, or information assets it holds and be able to manage and protect them appropriately. Important elements to consider when developing an asset and data management strategy are as follows:

- Inventory. (Do you know what assets you have & where they are?)
- Responsibility/ownership (Do you know who is responsible for each asset?)
- Importance. (Do you know how important each asset is in relation to other assets?)
- Establish acceptable-use rules for information and assets.
- Establish procedures for the labeling of physical and information assets.
- Establish return of asset procedures. (Do you have an employee exit procedure?)
- Protection. (Is each asset adequately protected according to how important it is?)

5.9.2 Level of assets

There are different levels at which asset units can be identified and managed, ranging from discrete equipment items or components to complex functional systems, networks, sites, or diverse portfolios. Many organizations identify assets as equipment units (sometimes referred to as "maintenance significant items" – the unit at which maintenance tasks or work orders are directed), whereas others use the term to describe functional systems or even integrated business units. It does not matter at what such level an asset unit is identified, provided that:

- The organization's goals and strategic priorities are directly reflected in the asset management plans.
- The asset life cycle costs, risks, and performance are considered and optimized. (This will usually require the definition of clear asset boundaries for measuring performance, life cycle expenditures, and attributing associated risks.)
- The aggregations of assets (through integrated asset systems) and contributions of value (as part of the organization's portfolio) are managed in a coordinated and consistent manner.
- All parts of the organization understand and use the same terminology in relation to the assets, their components, and their asset system groupings or aggregations.

This hierarchy brings challenges and opportunities at different levels. For example, discrete equipment items may have identifiable individual life cycles that can be optimized, whereas asset systems may have an indefinite horizon of required usage. Sustainability considerations should, therefore, be part of optimized decision making. A larger organization may also have a diverse portfolio of asset systems, each contributing to the overall goals of the organization but presenting widely different investment opportunities, performance challenges, and risks.

An integrated asset management system is therefore essential to coordinate and optimize the diversity and complexity of assets in line with the organization's objectives and priorities. The asset management focus will tend to differ at the various levels of asset integration in an organization. In Figure 5.3, the level of assets shows examples of priorities that might be evident at the different levels of asset integration and management

5.9.3 Asset management

Asset and data management is based on the idea that it is important to identify, track, classify, and assign ownership to the most important assets in your organization to ensure they are adequately protected. Tracking the inventory of IT hardware is the simplest example of asset management. Knowing what you have, where it lives, how important it is, and who's responsible for it are all important pieces of the puzzle. Similarly, an information asset is an item of value containing

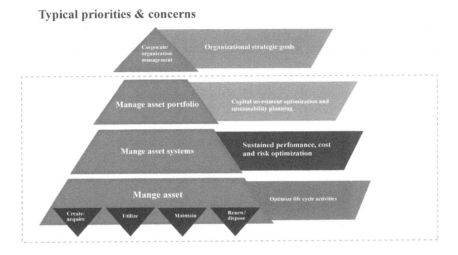

Figure 5.3 Level of assets.

information. Managing assets effectively for utilities is not optional these days. Across the globe, every society is faced with a significant asset management challenge:

- Emerging economies are trying to identify the lowest cost and highest return investments to achieve maximum immediate benefit.
- Rapidly developing countries are faced with understanding the life cycle costs of their infrastructure. More mature economies are trying to find ways of extending the life of their infrastructure and also meet major global challenges like climate change.

Asset management can be defined as systematic and coordinated activities and practices through which an organization optimally and sustainably manages its assets and asset systems, their associated performance, risks, and expenditures over its life cycles for the purpose of achieving its organizational strategic plan.

The same concepts of general asset management apply to the management of information assets (e.g., data). To be effective, an overall asset management strategy should include information assets, software assets, and information technology equipment. In addition, the people employed by an organization, as well as the organization's reputation, are also important assets not to be overlooked in an effective asset management strategy. An organization should be in a position to know what physical, environmental, or information assets it holds and be able to manage and protect them appropriately. Important elements to consider when developing an asset and data management strategy are:

- Inventory. (Do you know what assets you have & where they are?)
- Responsibility/ownership. (Do you know who is responsible for each asset?)
- Importance. (Do you know how important each asset is in relation to other assets?)
- Establish acceptable-use rules for information and assets.
- Establish procedures for the labeling of physical and information assets.
- Establish return of asset procedures. (Do you have an employee exit procedure?)
- Protection. (Is each asset adequately protected according to how important it is?)

Good asset management considers and optimizes the conflicting priorities of asset utilization and asset care, short-term performance opportunities and long-term sustainability, and between capital investments and subsequent operating costs, risks, and performance. "Life cycle" asset management is also more than simply the consideration of capital costs and operating costs over pre-determined asset "life" assumptions. Truly optimized, whole life asset management includes risk exposures and performance attributes and considers the asset's economic life as the result of an optimization process (depending upon the design, utilization, maintenance, obsolescence, and other factors).

Asset management is important because it can help organizations to:

1. Reduce the total costs of operating their assets.
2. Reduce the capital costs of investing in the asset base.
3. Improve the operating performance of their assets (reduce failure rates, increase availability, etc.).
4. Reduce the potential health impacts of operating the assets.
5. Reduce the safety risks of operating the assets.
6. Minimize the environmental impact of operating the assets.
7. Maintain and improve the reputation of the organization.
8. Improve the regulatory performance of the organization.
9. Reduce legal risks associated with operating assets.

The key to good asset management is that it optimizes these benefits. That means that asset management takes all of the above into account and determines the best blend of activities to achieve the best balance for all of the above for the benefit of the organization. Asset management is explicitly focused on helping organizations to achieve their defined objectives and to determine the optimal blend of activities based on these objectives.

5.9.4 The principles of asset management

Asset management is a holistic view and one that can unite different parts of an organization together in pursuit of shared strategic objectives. The key principles and attributes of successful asset management can be explained as follows (Figure 5.4).

- Holistic: Looking at the whole picture, i.e. the combined implications of managing all aspects (this includes the combination of different asset types, the functional interdependencies and contributions of assets within asset systems, and the different asset life cycle phases and corresponding activities), rather than a compartmentalized approach.
- Systematic: A methodical approach, promoting consistent, repeatable, and auditable decisions and actions.
- Systemic: Considering the assets in their asset system context and optimizing the asset systems value (including sustainable performance, cost, and risks) rather than optimizing individual assets in isolation.
- Risk-based: Focusing resources and expenditure, and setting priorities, appropriate to the identified risks and the associated cost/benefits.
- Optimal: Establishing the best value compromise between competing factors, such as performance, cost, and risk, associated with the assets over their life cycles.

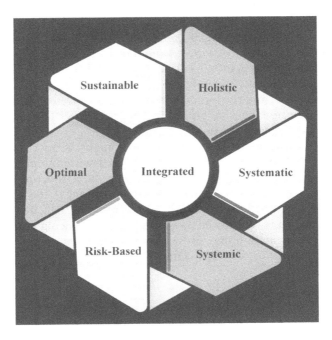

Figure 5.4 The principles of asset management.

- Sustainable: Considering the long-term consequences of short-term activities to ensure that adequate provision is made for future requirements and obligations (such as economic or environmental sustainability, system performance, societal responsibility, and other long-term objectives).
- Integrated: Recognizing that interdependencies and combined effects are vital to success. This requires a combination of the above attributes, coordinated to deliver a joined-up approach and net value.

A key principle in asset management is a LINE OF SIGHT that means:

- An approach within an organization that looks to line up the work that is done directly on assets with the objectives of that organization.
- A discipline which recognizes, accommodates, and aligns the risk of owning a particular asset with the goals of the organization that operates the asset.

Some examples

Eg.1. A good "asset management" decision might be to purchase an expensive, high-specification stainless steel piping system within an industrial process. While the initial cost is higher, the maintenance costs may be lower, the expected life is three times longer, the risk of disruptive failure may be lower, and therefore the risk to the organization from a performance, health & safety, and environmental perspective is consequently much lower. The total life cycle costs, therefore, may be lower, and the total risk to the organization through purchasing the more expensive piping system, therefore, represents a good asset management decision.

Eg.2. A poor asset management decision might be to reduce the frequency of maintenance activity on an asset without appreciating the full impact of doing so. While there may be a short-term financial benefit, the long-term cost to the organization, if the asset prematurely fails, might substantially outweigh this benefit. Of course, maintenance is recognized as a means of introducing failures, so the proper investigation may prove that reducing maintenance frequency is a net benefit to the organization!

5.9.5 Asset life cycle

Understanding that assets have a life cycle is a key concept within asset management and is therefore worthy of scrutiny. There are dozens of

Figure 5.5 Asset life cycle.

different ways of representing the life cycle, but Figure 5.5 captures a simple representation of it.

1. **Acquire**
   This covers everything that goes into planning, designing, and procuring an asset. Some life cycle diagrams capture planning as a separate function. Proper application of these activities ensures that the asset is fit for purpose.
2. **Commission**
   This covers the activities of installing/creating or building the asset and ensuring that it is fully functional. It is a recognized fact that there is a higher incidence of failure after the first installation/building of an asset. This is reflected in the need for the commissioning stage in the life cycle to oversee the initial operation of the assets.
3. **Operate**
   This is normally the bulk of the life cycle for an asset during which it provides the function for which it was designed. During this period, the asset should be subject to appropriate monitoring, maintenance, refurbishment, and potential upgrade to meet any change in condition or operational requirement. For many assets, this phase is decades-long. It may even be centuries. It is the phase that many engineers are most familiar with.
4. **Dispose**
   This is often the most overlooked phase. Assets can last beyond a human life time, and it can be difficult to consider asset disposal when it is so far into the future. Asset management teaches us that we ignore

any stage of the asset life cycle at our peril. This is a key period within an asset's life. With some assets, e.g. in the nuclear industry, this can be an extended and highly critical period. Key activities during this period include the effective removal of the asset from operation; the disposal or recycling of the asset or its components; and the feed into the planning for the replacement asset (if a replacement is required) to determine the operational requirements based on the effectiveness of operation and the failure modes encountered.

How to go about it?

1. Review potential sources of information assets. A holistic perspective that includes data centers, hardware, software, and data may require various sources including:
   - Organizational asset inventory reports from departments responsible for purchasing and equipment asset inventory.
   - Organizational information security risk assessments.
   - Business continuity and disaster recovery plans (a good source for critical systems).
   - Visit your organization's CIO and data center management and discuss what information resources are under their custody.
   - Visit major stakeholders (senior staff, administrative department heads, etc.,) and discuss what information systems and data their department handles.
   - Create a spreadsheet of the items.
   - List the assets for each category.
   - Define distinct categories for the types of assets in the organization (e.g., infrastructure, data center hardware, information systems/applications, data).
2. Record the physical location of the asset in your spreadsheet. You may want to divide them into local and hosted. Included under local are institutional brick and mortar physical locations such as classrooms, data centers, labs, or offices. For example, the location of collaborative research materials on a file share may be Primary Data Center X. Included under hosted are third-party vendor data centers and other remote locations not owned by the organization. For example, the location of the learning management system is the Vendor X data center located in address.
3. Identify and record in your spreadsheet the owners and custodians for each of the assets listed in your spreadsheet. Most of the time, the individuals responsible for the security of the asset and ensuring compliance are not the same as the individuals responsible for implementing security controls and day-to-day operations.
   - Review the federal or state laws, regulations, rules, or institutional policies that require protection of information resources.
   - Review your institution's Data Classification Policy.

- Determine if the organization's assets are classified in accordance with the Data Classification Policy.
- Create a simple classification schema (e.g., Public, Restricted, Confidential).
4. Create a criticality rating for the assets. For example (highest to lowest):
   - Critical is always available and protected.
   - Very important this asset is available and protected.
   - Important if this asset is available and protected.
   - Good if this asset is available with minimal protection.
5. Record in your spreadsheet the asset classification and/or criticality ranking.
   - Example 1: The LMS system has a rating of 2.
   - Example 2: Customer records are confidential and have a rating of 1.
6. Determine whether institutional assets are protected according to their classification and importance.

5.9.6 Seven steps to implement asset management

1. Developing policy: The asset management policy is the link between the organizational plan (that is, the top-level "business plan" in a company) and the asset management strategy. It is typically a set of principles or guidelines to steer asset management activity to achieve the organization's objectives. It specifically covers the "what" and the "why".
2. Developing strategy: The asset management strategy directs the organization's asset management activity; it will determine the high-level asset management objectives that are needed from the activity to deliver the organization's objectives; it will define the approach to planning that will be taken.
3. Asset management planning: Asset management planning looks at considering all the options for activities and investments going forward and then putting together a set of plans which describe what will be done when and by whom. The asset manager ensures that the plan delivers what is required of it by the strategy.
4. Delivering the plans: This is the bit where work is actually done on the assets, whether assessing or monitoring them, maintaining or repairing them, refurbishing or replacing them. This activity clearly needs to include the appropriate controls to ensure the work is done efficiently and that information gathered is fed back into the strategy and planning activities.
5. Developing people: This activity is specifically about developing the skills and competences of people to better deliver asset management activities. It spans from the boardroom to the toolbox and also

through the supply chain. As well as individual skills, it looks at the culture within an organization and how change can be managed to achieve optimal results for that organization.
6. Managing risk: Understanding risk is a critical concept in asset management and is a key function and area of competence. Its focus is on being able to assess the risk of action or inaction on the performance of assets in the context of the organization's corporate objectives.
7. Managing asset information: Collecting and collating the right information to inform asset management decisions is crucial to achieving asset management's success. Too much data confuses the picture and costs money to collect. Too little data results in decisions made in the dark (or at best the twilight!). Ensuring that the right people have the right information to make the best decisions is key.

5.10. RESPONSIBILITY FOR ASSETS

A.8.1 Responsibility for assets

Objective:
To identify organizational assets and define appropriate protection responsibilities.

A.8.1.1 Inventory of assets

Control:
Assets associated with information and information processing facilities should be identified and an inventory of these assets should be drawn up and maintained.

Implementation guidelines:
An organization should identify assets relevant in the life cycle of information and document their importance. The life cycle of information should include creation, processing, storage, transmission, deletion, and destruction. Documentation should be maintained in dedicated or existing inventories as appropriate. The asset inventory should be accurate, up to date, consistent, and aligned with other inventories. For each of the identified assets, ownership of the asset should be assigned and the classification should be identified. Inventories of assets help to ensure that effective protection takes place and may also be required for other purposes, such as health and safety, insurance, or financial (asset management) reasons. ISO provides examples of assets that might need to be considered by the organization when identifying assets. The process of compiling an inventory of assets is an important prerequisite of risk management.

A.8.1.2 Ownership of assets

Control:
Assets maintained in the inventory should be owned.

Implementation guidelines:
Individuals, as well as other entities having approved management responsibility for the asset life cycle, qualify to be assigned as asset owners. A process to ensure timely assignment of asset ownership is usually implemented. Ownership should be assigned when assets are created or when assets are transferred to the organization. The asset owner should be responsible for the proper management of an asset over the whole asset life cycle. The asset owner should:

- Ensure that assets are inventoried.
- Ensure that assets are appropriately classified and protected.
- Define and periodically review access restrictions and classifications to important assets, taking into account applicable access control policies.
- Ensure proper handling when the asset is deleted or destroyed.

The identified owner can be either an individual or an entity who has approved management responsibility for controlling the whole life cycle of an asset. The identified owner does not necessarily have any property rights to the asset. Routine tasks may be delegated e.g. to a custodian looking after the assets on a daily basis, but the responsibility remains with the owner. In complex information systems, it may be useful to designate groups of assets that act together to provide a particular service. In this case, the owner of this service is accountable for the delivery of the service, including the operation of its assets.

A.8.1.3. Acceptable use of assets

Control:
Rules for the acceptable use of information and of assets associated with information and information processing facilities should be identified, documented, and implemented.

Implementation guidelines:
Employees and external party users using or having access to the organization's assets should be made aware of the information security requirements of the organization's assets associated with information and information processing facilities and resources. They should be responsible for their use of any information processing resources and of any such use carried out under their responsibility.

A.8.1.4. Return of asset

Control:

All employees and external party users should return all of the organizational assets in their possession upon termination of their employment, contract, or agreement.

Implementation guidelines:

The termination process should be formalized to include the return of all previously issued physical and electronic assets owned by or entrusted to the organization. In cases where an employee or external party user purchases the organization's equipment or uses their own personal equipment, procedures should be followed to ensure that all relevant information is transferred to the organization and securely erased from the equipment. In cases where an employee or external party user has knowledge that is important to ongoing operations, that information should be documented and transferred to the organization. During the notice period of termination, the organization should control the unauthorized copying of relevant information (e.g. intellectual property) by terminated employees and contractors.

A.8.1.5. Responsibility for assets

In order to effectively manage an organization's assets, you must first understand what assets you have and where your organization keeps them. Some asset examples are IT hardware, software, data, system documentation, and storage media. Supporting assets such as data center air systems, UPSs, and services should be included in the inventory. All assets should be accounted for and have an owner. If improperly managed, assets can become liabilities.

- Categorize your assets: Begin by defining distinct categories of the types of assets in the organization. Each category should have its own inventory or classification structure based on the assets that category may contain (Category: Data Center Hardware).
- Create a list of assets for each category: Creating a list of an institution's assets and their corresponding locations is the beginning of your inventory. Often, the process of doing so helps identify additional assets that previously had not been considered (Category: Data Center Hardware; Asset: Core Network Switches).
- Add a location for each asset: The location could be a brick and mortar physical location such as a classroom, data center, or office. It could also be collaborative research materials on a file share or financial information stored in a database (Category: Data Center Hardware; Asset: Core Network Switches; Location: Room no 001).

Because assets can be many things and serve multiple functions, there will likely be more than one inventory process or system used to capture the range of assets that exist in an organization. Make sure you connect with other areas to see what form of hardware inventory already exists. Don't start from zero. Each inventory system should not unnecessarily duplicate other inventories that may exist.

Asset responsibility/ownership

Once you have begun to capture an inventory of the potential assets and their locations, start identifying the responsible person for each asset. An owner is a person, or persons or department that has been given formal responsibility for the security of an asset. The owner is responsible for securing the asset during the life cycle of the asset. At this juncture in the exercise, it is important to understand the distinction between the terms "owner" and "custodian" of assets.

The custodian is responsible for ensuring that the asset is managed appropriately over its lifecycle, in accordance with rules set by the asset owner. The custodian is often a subject matter expert (SME) or "owner" of the business process for a particular information asset. An owner of an information asset, Data Owners, will have direct operational responsibility for the management of one or more types of data. Think of it in terms of an information security department. You have the "owner", the person responsible for interpreting and assuring compliance. That would be the Director or CISO. Then there is the custodian, the person responsible for the day-to-day operations and management of the tools and processes that protect the information assets.

Identifying the owners will help determine who will be responsible for carrying out protective measures and responding to situations where assets may have been compromised. You will also quickly realize when it isn't clear who the appropriate responsible party is or when shared responsibility may be an issue.

(Category: Data Center Hardware; Asset: Core Network Switches; Location: Room no 01; Owner: Director XYZ)

The owner of the assets should be able to identify acceptable uses or provide information on which policy governs its acceptable use. Work with the responsible owner, if need be, on acceptable uses. The acceptable uses should include items such as who assumes the risk of loss, gives access to the asset, and how a critical asset is kept functional during or after a loss. Policies governing the use, preservation, and destruction of hardware may originate from your asset management office. Many organizations also find it helpful to document expectations for the acceptable and responsible use of information technology assets in an Acceptable and Responsible Use Policies.

Physical and environmental asset importance

All assets add value to an organization. However, not all assets are created equal. Gaining a clear understanding of the relative importance of each asset when compared to other organizational assets is an essential step if you are to adequately protect your assets. The importance of an asset can be measured by its business value and security classification or label. Create a rating system for the asset. It can be as simple as (highest to lowest)

- Critical is always available and protected.
- Very important this asset is available and protected.
- Important if this asset is available and protected.
- Good if this asset is available with minimal protection.

Building on the previous example and adding a rating system, it would look like (Category: Data Center Hardware; Asset: Core Network Switches; Location: Room no 01; Owner: Director XYZ; Rate: 1 (Critical)).

A computer kept in a cafeteria for the purpose of recreation may have a lower score given it is good that the asset is available. The computer kept in the finance department may be protected with anti-virus and firewall.

Acceptable use of assets associated with information

After going through the asset inventory, categorization, and ownership identification, ensure there are documented policies regarding the acceptable use of assets. Define, and document, the rules that clarify the acceptable uses of assets associated with information and information processing facilities. It is important, once the rules are clarified, that appropriate controls are implemented and the security requirements are communicated. Target the communication of security requirements to employees and, if appropriate, third parties who may use these assets. Accountability is key. Asset owners should be responsible and accountable, even if the owner has delegated responsibility, for their use of facilities and resources.

An example of an acceptable-use policy

A. General use and ownership
  a. Proprietary information stored on electronic and computing devices whether owned or leased by, the employee or a third party, remains his/her sole property. You must ensure through legal or technical means that proprietary information is protected in accordance with the Data Protection Standard.
  b. You have a responsibility to promptly report the theft, loss, or unauthorized disclosure of proprietary information.

c. You may access, use, or share proprietary information only to the extent it is authorized and necessary to fulfill your assigned job duties.
d. Employees are responsible for exercising good judgment regarding the reasonableness of personal use. Individual departments are responsible for creating guidelines concerning personal use of Internet/Intranet/Extranet systems. In the absence of such policies, employees should be guided by departmental policies on personal use, and if there is any uncertainty, employees should consult their supervisor or manager.
e. For security and network maintenance purposes, authorized individuals within may monitor equipment, systems, and network traffic at any time, per the organization's audit policy.
f. Reserves the right to audit networks and systems on a periodic basis to ensure compliance with this policy.

B. Security and proprietary Information
   a. All mobile and computing devices that connect to the internal network must comply with the minimum access policy.
   b. System-level and user-level passwords must comply with the password policy. Providing access to another individual, either deliberately or through failure to secure its access, is prohibited.
   c. All computing devices must be secured with a password-protected screensaver with the automatic activation feature set to 10 minutes or less. You must lock the screen or log off when the device is unattended.
   d. Postings by employees from an email address to newsgroups should contain a disclaimer stating that the opinions expressed are strictly their own and not necessarily those of unless posting is in the course of business duties.
   e. Employees must use extreme caution when opening email attachments received from unknown senders, which may contain malware.

C. **Unacceptable use**
   The following activities are, in general, prohibited. Employees may be exempted from these restrictions during the course of their legitimate job responsibilities (e.g., systems administration staff may have a need to disable the network access of a host if that host is disrupting production services).
   Under no circumstances is an employee authorized to engage in any activity that is illegal under local, state, federal, or international law while utilizing owned resources.

The lists below are by no means exhaustive but attempt to provide a framework for activities that fall into the category of unacceptable use.

a. **System and network activities**

The following activities are strictly prohibited, with no exceptions:

I. Violations of the rights of any person or company protected by copyright, trade secret, patent or other intellectual property, or similar laws or regulations, including, but not limited to, the installation or distribution of "pirated" or other software products that are not appropriately licensed for use by.

II. Unauthorized copying of copyrighted material including, but not limited to, digitization and distribution of photographs from magazines, books or other copyrighted sources, copyrighted music, and the installation of any copyrighted software for which or the end-user does not have an active license is strictly prohibited.

III. Accessing data, a server, or an account for any purpose other than conducting business, even if you have authorized access, is prohibited.

IV. Exporting software, technical information, encryption software, or technology, in violation of international or regional export control laws, is illegal. The appropriate management should be consulted prior to the export of any material that is in question.

V. Introduction of malicious programs into the network or server (e.g., viruses, worms, Trojan horses, and email bombs).

VI. Revealing your account password to others or allowing the use of your account by others. This includes family and other household members when work is being done at home.

VII. Using a computing asset to actively engage in procuring or transmitting material that is in violation of sexual harassment or hostile workplace laws in the user's local jurisdiction.

VIII. Making fraudulent offers of products, items, or services originating from any account.

IX. Making statements about warranty, expressly or implied, unless it is a part of normal job duties.

X. Effecting security breaches or disruptions of network communication. Security breaches include, but are not limited to, accessing data of which the employee is not an intended recipient or logging into a server or account that the employee is not expressly authorized to access unless these duties are within the scope of regular duties. For purposes of this section, "disruption" includes, but is not limited to, network sniffing, pinged floods, packet spoofing, denial of service, and forged routing information for malicious purposes.

XI. Port scanning or security scanning is expressly prohibited unless prior notification to information security is made.

XII. Executing any form of network monitoring which will intercept data not intended for the employee's host, unless this activity is a part of the employee's normal job/duty.
XIII. Circumventing user authentication or security of any host, network, or account.
XIV. Introducing honeypots, honeynets, or similar technology on the network.
XV. Interfering with or denying service to any user other than the employee's host (for example, denial of service attack).
XVI. Using any program/script/command, or sending messages of any kind, with the intent to interfere with, or disable, a user's terminal session, via any means, locally or via the Internet/Intranet/Extranet.
XVII. Providing information about, or lists of, employees to parties outside.

b. **Email and communication activities**

When using company resources to access and use the Internet, users must realize they represent the company. Whenever employees state an affiliation to the company, they must also clearly indicate that "the opinions expressed are my own and not necessarily those of the company". Questions may be addressed to the IT department.

I. Sending unsolicited email messages, including the sending of "junk mail" or other advertising material to individuals who did not specifically request such material (email spam).
II. Any form of harassment via email, telephone, or paging, whether through language, frequency, or size of messages.
III. Unauthorized use, or forging, of email header information.
IV. Solicitation of email for any other email address, other than that of the poster's account, with the intent to harass or to collect replies.
V. Creating or forwarding "chain letters", "Ponzi", or other "pyramid" schemes of any type.
VI. Use of unsolicited email originating from within 's networks of other Internet/Intranet/Extranet service providers on behalf of, or to advertise, any service hosted by or connected via 's network.
VII. Posting the same or similar non-business-related messages to large numbers of Usenet newsgroups (newsgroup spam).

c. **Blogging and social media**

I. Blogging by employees, whether using 's property and systems or personal computer systems, is also subject to the terms and restrictions set forth in this policy. Limited and occasional use of 's systems to engage in blogging is acceptable, provided that it is done in a professional and responsible manner, does not otherwise violate 's policy, is not detrimental to 's best interests, and does not interfere with an employee's regular work duties. Blogging from 's systems is also subject to monitoring.

II. 's Confidential Information policy also applies to the blog. As such, employees are prohibited from revealing any confidential or proprietary information, trade secrets, or any other material covered by 's Confidential Information policy when engaged in blogging.
III. Employees shall not engage in any blogging that may harm or tarnish the image, reputation, and/or goodwill of and/or any of its employees. Employees are also prohibited from making any discriminatory, disparaging, defamatory, or harassing comments when blogging or otherwise engaging in any conduct prohibited by 's Non-Discrimination and Anti-Harassment policy.
IV. Employees may also not attribute personal statements, opinions, or beliefs to when engaged in blogging. If an employee is expressing his or her beliefs and/or opinions in blogs, the employee may not, expressly or implicitly, represent himself or herself as an employee or representative. Employees assume any and all risks associated with blogging.
V. Apart from following all laws pertaining to the handling and disclosure of copyrighted or export controlled materials, 's trademarks, logos, and any other intellectual property may also not be used in connection with any blogging activity

d. **Policy compliance**
I. Compliance measurement: The Infosec team will verify compliance with this policy through various methods, including but not limited to, business tool reports, internal and external audits, and feedback to the policy owner.
II. Exceptions: Any exception to the policy must be approved by the Infosec team in advance.
III. Non-compliance: An employee found to have violated this policy may be subject to disciplinary action, up to and including termination of employment.

Return of assets

It is critical that organizations protect their information on the equipment of employees when their employment is terminated. Make sure all relevant information that will be needed by the institution is preserved, but all information on the asset is erased. Develop an employee exit checklist that addresses the return of all institutional assets, physical or information, before the employee's last day. There are, of course, emergency situations dealing with immediate termination that may not lend themselves to a measured checklist. Create a simple checklist for those instances as well. Get to know a resource in your HR area and work with that resource to incorporate physical and electronic assets at termination.

As stated before, assets can be a variety of items. Employee knowledge is also an information asset to the organization. Preserve their relevant knowledge, document before the individual leaves the institution, and ensure that knowledge is in the organization's possession. Once again, use the checklist to incorporate this aspect of asset return. A sample may include:

- The employee has returned all computing equipment to IT.
- IT will preserve the information on the equipment by copying to an external drive or employee group shared file server. Preserved.
- Information will be given to the employee's supervisor.
- The employee has transferred all institutional information from his/her personal equipment and given that to their supervisor.
- Employee rights to information assets have been terminated as of this date.
- Employee knowledge transfer has occurred.

Don't forget about the contractors, consultants, or any other external third party upon termination of the contract or agreement. The same rules apply. You may wish to have a separate asset security checklist for all external agents and ensure this information is part of their contract or agreement.

5.11. INFORMATION CLASSIFICATION

A.8.2 Information classification

Objective:
To ensure that information receives an appropriate level of protection in accordance with its importance to the organization.

A.8.2.1 *Classification of information*

Control:
Information should be classified in terms of legal requirements. value, criticality, and sensitivity to unauthorized disclosure or modification.

Implementation guidelines:
Classifications and associated protective controls for information should take account of business needs for sharing or restricting information, as well as legal requirements. Assets other than information can also be classified in conformance with the classification of information that is stored in, processed by, or otherwise handled or protected by the asset. Owners of information assets should be accountable for their classification. The classification scheme should include conventions for classification and criteria for

review of the classification over time. The level of protection in the scheme should be assessed by analyzing confidentiality, integrity, and availability, and any other requirements for the information considered. The scheme should be aligned with the access control policy. Each level should be given a name that makes sense in the context of the classification scheme's application. The scheme should be consistent across the whole organization so that everyone will classify information and related assets in the same way, have a common understanding of protection requirements, and apply for the appropriate protection.

The classification should be included in the organization's processes and be consistent and coherent across the organization. Results of classification should indicate the value of assets depending on their sensitivity and criticality to the organization, e.g. in terms of confidentiality, integrity, and availability. Results of classification should be updated in accordance with changes in their value, sensitivity, and criticality through their life cycle. Classification provides people who deal with information with a concise indication of how to handle and protect it. Creating groups of information with similar protection needs and specifying information security procedures that apply to all the information in each group facilitates this. This approach reduces the need for case-by-case risk assessment and custom design of controls.

Information can cease to be sensitive or critical after a certain period of time, for example, when the information has been made public. These aspects should be taken into account as over-classification can lead to the implementation of unnecessary controls resulting in additional expense or on the contrary, under-classification can endanger the achievement of business objectives. An example of an information confidentiality classification scheme could be based on four levels as follows:

a. Disclosure causes no harm.
b. Disclosure causes minor embarrassment or minor operational inconvenience.
c. Disclosure has a significant short-term impact on operations or tactical objectives.
d. Disclosure has a serious impact on long-term strategic objectives or puts the survival of the organization at risk.

A.8.2.2 Labeling of information control

An appropriate set of procedures for information labeling should be developed and implemented in accordance with the information classification scheme adopted by the organization.

Implementation guidelines:

Procedures for information labeling need to cover information and its related assets in physical and electronic formats. The labeling should reflect the classification scheme established in 8.2.1. The labels should be easily recognizable. The procedures should give guidance on where and how labels are attached in consideration of how the information is accessed or the assets are handled depending on the types of media. The procedures can define cases where labeling is omitted, e.g. labeling of non-confidential information to reduce workloads.

Employees and contractors should be made aware of labeling procedures. The output from systems containing information that is classified as being sensitive or critical should carry an appropriate classification label. Labeling of classified information is a key requirement for information sharing arrangements. Physical labels and metadata are a common form of labeling. Labeling information and its related assets can sometimes have negative effects. Classified assets are easier to identify and accordingly to steal by insiders or external attackers.

A.8.2.3 Handling of assets

Control:
Procedures for handling assets should be developed and implemented in accordance with the information classification scheme adopted by the organization.

Implementation guidelines:
Procedures should be drawn up for handling processing, storing, and communicating information consistent with its classification. The following items should be considered:

1. Access restrictions supporting the protection requirements for each level of classification.
2. Maintenance of a formal record of the authorized recipients of assets.
3. Protection of temporary or permanent copies of information to a level consistent with the protection of the original information.
4. Storage of IT assets in accordance with manufacturers' specifications.
5. Clear marking of all copies of media for the attention of the authorized recipient.

The classification scheme used within the organization may not be equivalent to the schemes used by other organizations. Even if the names for levels are similar in addition, information moving between organizations can vary in classification depending on its context in each organization, even if their classification schemes are identical. Agreements with other organizations that include information sharing should include procedures to identify the classification of that information and to interpret the classification labels from other organizations.

Data protection and privacy of personal information (records management)

The valuable data every organization needs to be protected commensurate with how it is classified. Customers, employees, and vendors entrust the organization with a given data set and there is an implied bargain that the data so entrusted will be protected from any use or disclosure other than as agreed to when the data was given. To do this, each organization has to govern the data it uses so that it will be received, made, used, stored, shared, or destroyed in a purposeful manner that recognizes the pact to protect data in its daily mission. Areas to consider in a data governance program include:

- **Sensitivity level.** An organization should be classifying data as to sensitivity to assure that proper security protection is in place appropriate with the given data set.
- **Retention period.** Consistent with records management practices, an organization needs to be aware of the period in which data is to be retained, to assure that data's availability and integrity for that retention period.
- **Data utilization.** In every part of an organization that controls a given data set, appropriate procedures for how that data is utilized must be established. This includes access restrictions, proper handling, logging, and auditing.
- **Data back-up.** How an organization creates back-up copies of data and software is a critical element. Procedures need to be in the place that memorialize and verify the implementation and inventory of back-up copies.
- **Management of storage media.** Processes to ensure proper management of storage media, including restrictions of types of media, audit trails for movement of media, secure disposal of media no longer in use, and redundant storage.
- Electronic Data Transfers.
- Disposal of Media.

Information asset importance

Information assets may not be equally important, equally sensitive, or confidential in nature, nor require the same care in handling. One common method of ascertaining the importance of assets is data classification. Information assets should be classified according to their need for security protection and labeled accordingly. To begin to start with federal or state laws, regulations, rules, or institutional policies that require certain information assets to be protected, pick a classification metric. Keep it simple. You may want to use something like (lowest to highest) public, restricted, confidential.

Asset protection

Different assets have different impacts on the continuity and reputation of the organization. Once you have determined the importance of your various organizational assets, you can begin the process of determining how best to protect them. Many methods are employed to protect assets, ranging from policies to technical security controls. Additionally, assets must be protected throughout their life cycle, from creation or purchase through final disposal or long-term storage. Protection measures range from addressing purchasing controls to managing access by appropriate personnel to ensuring adequate physical security for assets throughout their lifetime.

Some organization has established data stewardship policies to help ensure responsibilities for protecting data are effectively accomplished. Other organizations conduct regular security assessments of assets considered to be critical for the functioning of an organization. They may also address asset protection through physical security measures or through background checks for newly hired and continuing personnel.

Labeling of information

DO YOU HAVE YOUR INFORMATION AND PHYSICAL ASSETS LABELED?

Your organization may already have property control of assets where items over a certain dollar amount are automatically tagged with a unique, usually numeric, identifier by property control. If not, create one yourself. Use your newly created inventory of assets to assign a unique identifier to each one. Prepare labels that are easy to recognize and sturdy and attach them to a visible place on the equipment. Make sure you clarify when labels should not be used on equipment. This could be based on the dollar amount or the level of risk you've assigned to the asset. Information needs labeling as well. Develop your information labeling procedures based on the data classification schema you developed previously. Metadata is a common type of information label. Do be careful how you manage the information you may have labeled as restricted or confidential. Because of the labeling, be careful how you manage restricted/sensitive or confidential information. It is much easier to steal or misuse when the assets are easy to identify.

Handling of assets

IS INFORMATION BEING HANDLED AND PROTECTED ACCORDING TO ITS CLASSIFICATION?

Now that you have your assets identified, classified, and labeled, you will need to develop procedures for handling assets associated with your information and information processing facilities. It is important that your asset handling procedures respect and reflect how you classified it. Ensure that

- Information is handled and protected according to its classification. This includes sharing with external entities.
- There are procedures to control classified information. Clarify how yours and perhaps others' classifications should be interpreted.
- Information is stored, processed, transmitted, and copied according to its classification. Copies should get the same protections.
- Access restrictions are designed for each level of classification. Restrictions must meet protection requirements.
- There is a formal record of the authorized recipients of the assets. Specify who the authorized recipient should be. Label media copies appropriately.

All of the above bullet points can be incorporated into one procedural access-handling document. Remember, keep it simple so others will be able to understand and comply with the requirements. Hold a session with your information and physical asset owners so they can help you define the requirements. It's important everyone feels ownership of this process.

5.12. MEDIA HANDLING

A.8.3 Media handling

Objective:
To prevent unauthorized disclosure, modification, removal, or destruction of information stored on media.

A.8.3.1 Management of removable media

Control:
Procedures should be implemented for the management of removable media in accordance with the classification scheme adopted by the organization.

Implementation guidelines:
The following guidelines for the management of removable media should be considered:

a. If no longer required, the contents of any re-usable media that are to be removed from the organization should be made unrecoverable.
b. Where necessary and practical, authorization should be required for media removed from the organization, and a record of such removals should be kept in order to maintain an audit trail.

Asset Management

c. All media should be stored in a safe, secure environment, in accordance with manufacturers' specifications.
d. If data confidentiality or integrity are important considerations, cryptographic techniques should be used to protect data on removable media.
e. To mitigate the risk of media degrading while stored data are still needed, the data should be transferred to fresh media before becoming unreadable.
f. Multiple copies of valuable data should be stored on separate media to further reduce the risk of coincidental data damage or loss.
g. Registration of removable media should be considered to limit the opportunity for data loss.
h. Removable media drives should only be enabled if there is a business reason for doing so.
i. Where there is a need for muse removable media, the transfer of information to such media should be monitored.
j. Procedures and authorization levels should be documented.

A.8.3.2 Disposal of media

Control:
Media should be disposed of securely when no longer required, using formal procedures.

Implementation guidelines:
Formal procedures for the secure disposal of media should be established to minimize the risk of confidential information leakage to unauthorized persons. The procedures for the secure disposal of media containing confidential information should be proportional to the sensitivity of that information. The following items should be considered:

a. Media containing confidential information should be stored and disposed of securely, e.g. by incineration or shredding or erasure of data for use by another application within the organization.
b. Procedures should be in place to identify the items that might require secure disposal.
c. It may be easier to arrange for all media items to be collected and disposed of securely, rather than attempting to separate the sensitive items.
d. Many organizations offer collection and disposal services for media; care should be taken in selecting a suitable external party with adequate controls and experience;
e. Disposal of sensitive items should be logged in order to maintain an audit trail.

When accumulating media for disposal, consideration should be given to the aggregation effect, which can cause a large quantity of non-sensitive information to become sensitive. Damaged devices containing sensitive data may require a risk assessment to determine whether the items should be physically destroyed rather than sent for repair or discarded.

A.8.3.3 Physical media transfer

Control:
Media containing information should be protected against unauthorized access, misuse, or corruption during transportation.

Implementation guidelines:
The following guidelines should be considered to protect media containing the information being transported:

a. Reliable transport or couriers should be used.
b. A list of authorized couriers should be agreed upon with the management.
c. Procedures to verify the identification of couriers should be developed.
d. Packaging should be sufficient to protect the contents from any physical damage likely to arise during transit and in accordance with any manufacturers' specifications, for example protecting against any environmental factors that may reduce the media's restoration effectiveness such as exposure to heat, moisture, or electromagnetic fields.
e. Logs should be kept, identifying the content of the media, the protection applied as well as recording the times of transfer to the transit custodians and receipt at the destination.

Information can be vulnerable to unauthorized access, misuse, or corruption during physical transport, for instance when sending media via the postal service or via courier. In this control, the media include paper documents. When confidential information on media is not encrypted, additional physical protection of the media should be considered.

Management of removable media

Integrate necessary controls to manage media items, whether tapes, disks, flash disks, or removable hard drives, CDs, DVDs, or printed media, to ensure the integrity and confidentiality of university data. Guidelines should be developed and implemented to ensure that media are used, maintained, and transported in a safe and controlled manner. Handling and storage should correspond with the sensitivity of the information in the media. Procedures to erase media if no longer needed, to ensure information is not leaked, are also important.

Disposal

Procedures for handling classified information should cover the appropriate means of its destruction and disposal. Serious breaches of confidentiality occur when apparently worthless disks, tapes, or paper files are dumped without proper regard for their destruction.

Information handling procedures

Procedures for handling and storage of sensitive information, together with audit trails and records, are important. Accountability should be introduced and data classification and risk assessments are performed, to ensure that necessary controls are applied to protect sensitive data. Appropriate access controls should be implemented to protect information from unauthorized disclosure or usage. Systems are also vulnerable to the unauthorized use of system documentation; much of this type of information should be regarded and handled as confidential. Security procedures, operating manuals, and operation records all come into this category.

5.13. BYOD

5.13.1 What are the types of BYOD?

BYOD stands for bring your own device. It's an IT policy that allows, and sometimes encourages, employees to access enterprise data and systems using personal mobile devices such as smartphones, tablets, and laptops.

There are four basic options or access levels to BYOD:

- Unlimited access for personal devices.
- Access only to non-sensitive systems and data.
- Access, but with IT control over personal devices, apps, and stored data.
- Access, but preventing local storage of data on personal devices.

5.13.2 Why is BYOD important?

BYOD policy is important because it helps organizations strike a balance between improved productivity and managed risk.

BYOD as a work practice appears inevitable. Forbes reports that 60 percent of millennial workers and 50 percent of workers over 30 think the tools they bring from their non-working life are more effective and productive than those that come from work. What is termed the BYOD market is expected to hit almost $367 billion by 2022, up from $30 billion in 2014, Forbes also points out.

Security risks and additional complexity persist. But which is riskier asks IBM security expert Jeff Crume?

"Letting employees who may know little about threats or mitigation strategies sort out what the most appropriate defenses are, install the proper tools, configure them for optimal usability/security, and maintain all this in the face of an ever-changing backdrop of newly discovered vulnerabilities and attack types.

Letting subject matter experts chart the course and enable members of the user community to focus on their daily jobs".

Most IT organizations, sensibly, are going with option B – which makes BYOD an inevitability for them and their teams. As such, BYOD becomes more than letting somebody from finance work on quarterly results from their tablet at home. It elevates BYOD to an IT imperative challenged with enabling a mobile workforce while mitigating the risks.

5.13.3 Benefits of BYOD improve productivity

Employees are more comfortable and proficient with their own devices. They are also more apt to adopt leading-edge features, and they don't have to manage two devices.

Boost employee satisfaction

BYOD can also boost employee satisfaction by letting employees use the devices they choose and prefer.

Cut enterprise costs

BYOD can also help cut costs by shifting device costs to the user and away from the IT budget.

Attract new hires

There is enough to acclimate to when entering a new organization that using your own device at work can be beneficial for that initial productivity. It is also a selling point for both Android and Apple users to know that they will not be required to use a different device type or learn a new mobile OS.

5.13.4 Risks of BYOD

As users potentially mix their personal and professional lives on their devices, they can unwittingly expose sensitive data or create vulnerabilities to malware (malicious software) and destructive cyberattack.

An initial concern with BYOD was the loss of the actual personal device and the sensitive or proprietary data on it. According to a 2014 study, the ability to remotely wipe lost devices was the policy most enforced by organizations interviewed.

Unfortunately, cyber attackers are opportunistic and soon found vulnerabilities through mobile applications and operating systems. By 2015, mobile devices monitored by IBM Trustee showed an active malware infection rate equal to PCs. In addition to managing security threats, BYOD can also mean additional tasks and responsibilities for IT departments – for devices they do not own or officially control. This brings a new level of complexity to IT functions and concerns such as help desk support, regulatory compliance, provisioning, asset management, data privacy and more.

5.13.5 Keys to effective BYOD

For BYOD to be effective, policies need to be developed and deployed that support productivity, enforce security, and operate efficiently to meet business requirements.

There are software technologies that can help.

Enterprise mobility management (EMM) and *mobile device management (MDM)* solutions can help enroll users and enforce secure BYOD policies, such as identity management and authentication procedures. *Unified Endpoint Management (UEM)* has evolved to enable IT organizations to consolidate disparate programs for provisioning, securing, and supporting mobile devices into a single solution. UEM can survey and report on devices enrolled with an IT department and provide a single, dashboard view of their management. UEM solutions are also incorporating artificial intelligence (AI) technologies to surface anomalies in vast amounts of data and recommend actions to remediate malware and other security incidents.

"As enterprises undertake or expand mobile deployments, they will need to get their arms around which deployment choices and which suppliers work best for them. Most enterprises will not be able to keep pace, nor will they have the technology, staffing, and processes in place or the ability to capitalize on mobile assets to deploy and optimize a mobile strategy to its full potential. As a result, IDC believes that the need for external IT services that can help enterprises plan, build, integrate, and manage their mobility initiatives will grow in importance".

5.13.6 Guidelines to help plan and implement effective BYOD

- Create policy before procuring technology by looking at key questions and factors – and considering all the key mobile players.
- What devices will be supported – or not? Who will pay for the data plan? What are, if any, the compliance issues of that data? What are the privacy implications for the company and employees? Each organization will have its own questions and ensuing policy decisions.

- Find the devices that are accessing corporate resources with tools that can communicate continuously with an email environment and detect all devices connected to the network.
- Enrollment should be simple and protected and configure the device at the same time. In a perfect scenario, users follow an email. link or text to a profile on their device, including an Acceptable Usage Agreement or AUA for network access.
- Configure devices over-the-air to avoid further help desk requests. All profiles, credentials, and settings should be delivered to the device. This is also an opportunity to create policies to restrict access to certain applications and generate warnings about data limits.
- Help users help themselves by enabling self-service for functions such as PINs, passwords, geo-location, and device wiping.
- Keep personal information private by communicating privacy policies to employees and providing settings capabilities to disable app inventory reporting and location services.
- Separate personal information from corporate data by making sure an MDM solution can selectively wipe corporate data should an employee leave and provide the option to wipe the entire device should it be lost.
- Manage data usage by setting roaming and in-network megabit limits and customizing the billing day to create notifications based on percentage used.
- Continually monitor and address devices for non-compliance by looking for activity like "jailbreaking", where a user may attempt to get paid apps for free; use SMS to notify of any non-compliance before hitting the wipe button; and work.
- Enjoy the return on investment (ROI) from BYOD by examining costs associated with shifting mobile device costs to employees such as device purchase, subsidized data plans – and include the costs of mobile device management solutions and services.

BIBLIOGRAPHY

1. Alebrahim, A., Hatebur, D. and Goeke, L., 2014. Pattern-based and ISO 27001 compliant risk analysis for cloud systems. In 2014 IEEE 1st International Workshop on Evolving Security and Privacy Requirements Engineering (ESPRE) (pp.42–47).
2. Calomiris, C.W., Himmelberg, C.P. and Wachtel, P., 1995. Commercial paper, corporate finance, and the business cycle: A microeconomic perspective. *Carnegie-Rochester Conference Series on Public Policy*, 42, pp.203–250.
3. Fanchi, J., 2010. *Integrated reservoir asset management: principles and best practices*. Gulf Professional Publishing.

4. Amadi-Echendu, J.E. et al., 2010. What is engineering asset management? In *Definitions, concepts and scope of engineering asset management* (pp.3–16). Springer.
5. Safa, N.S., Von Solms, R. and Furnell, S., 2016. Information security policy compliance model in organizations. *Computers & Security*, 56, pp.70–82.
6. Pahnila, S., Siponen, M. and Mahmood, A., 2007. Employees' behavior towards IS security policy compliance. In 2007 40th Annual Hawaii International Conference on System Sciences (HICSS'07) (pp.156b–156b).
7. Miller, K.W., Voas, J. and Hurlburt, G.F., 2012. BYOD: Security and privacy considerations. *It Professional*, 14(5), pp.53–55.
8. Garba, A.B., Armarego, J., Murray, D. and Kenworthy, W., 2015. Review of the information security and privacy challenges in Bring Your Own Device (BYOD) environments. *Journal of Information Privacy and Security*, 11(1), pp.38–54.

Index

A

Access control, 6, 7, 122, 131, 145, 154, 161
Active attack, 2–5
Additional regulations and frameworks, 20
Agreement, 19, 21, 29, 30, 39, 95, 124, 132, 146, 153, 155, 164
Allocate Resources and Train the Staff, 123
Apache, 25, 26, 69–71
Architecture, 1
Assess the Risks, 117, 129
Asset, 88, 116–119, 121, 127, 128, 131, 134–148, 150, 152, 153, 156–158, 163, 164
Asset Inventory, 128, 142, 144, 148
Asset Management, 127, 131, 133–141, 143–145, 147, 149, 151, 153, 155, 157, 159, 161, 163, 164
Asset Owner, 127, 128, 145, 148, 158
Attack, 1–6, 9, 11, 30, 107, 151, 155, 162
Attackers, 107, 155, 163
Audit, 8, 19, 20, 22, 33–52, 57–62, 64–70, 74–77, 79–97, 99–103, 105, 109, 123–126, 130, 133, 149, 152, 156, 158, 159, 161
Auditability, 10
Auditable, 59, 139
Audit competence and evaluation methods, 33
Auditing, 33, 35, 38, 43, 46, 53, 62, 75, 85, 92, 94, 97, 99, 105, 126, 156
Audit of individuals, 39
Audit of sole trader's books of accounts, 39
Auditor, 33–37, 39–51, 59, 61, 63–66, 85–99, 108, 125, 126, 130, 133
Auditor Quality and Selection, 63
Audit Planning and Preparation, 33, 35, 37, 39, 41, 43, 45, 47, 49, 51, 53, 55, 57, 59, 61
Audit Principles, 34
Audit Responsibilities, 46
Audit Script, 66–70, 74, 76, 79–81, 83, 84
Audit Stages, 85
Audit Team Meeting, 94
Audit Techniques, 63, 65, 67, 69, 71, 73, 75, 77, 79, 81, 83, 85–89, 91, 93, 95, 97, 99, 101, 103, 105
Audit Techniques and Collecting Evidence, 63, 65, 67, 69, 71, 73, 75, 77, 79, 81, 83, 85, 87, 89, 91, 93, 95, 97, 99, 101, 103, 105
Audit Time and Process Flow, 52
Authentication, 5–7, 151, 163
Availability, 6, 10–13, 43, 65, 66, 87, 107, 109, 114–117, 127, 138, 154, 156

B

Brute-force, 107
Business, 10, 11, 14, 15, 18–21, 33, 34, 36, 39, 40, 42, 43, 45, 47, 57–59, 65, 66, 88, 90–93, 95, 100, 108, 109, 112, 114, 117–119, 121–124, 126, 132, 134, 136, 142, 143, 147–154, 159, 163, 164
BYOD, 123, 161–164

C

Case study, 55, 126
Category, 19, 20, 28, 128, 142, 146–148, 150, 161
Chief Financial Officer, 38
Cipher, 9, 10
Cipher text, 9, 10
Code, 6, 8, 9, 29, 47, 50, 56, 67, 68, 103, 107, 108, 110
Collecting Evidence through Questions, 88
Communication, 2, 4–6, 14, 47, 65, 76, 95, 110, 131, 148, 150, 151
Compliance, 10, 11, 13, 19, 21, 28–30, 35, 50, 58–60, 65, 68, 91, 92, 95, 109, 132, 142, 147, 149, 152, 163, 164
Compliance vs. Conformance, 28
Computer, 8, 10, 13–19, 22–25, 27, 30, 105, 126, 128, 148, 151
Concept(s), 11–13, 99, 100, 107, 108, 127, 134, 137, 140, 144
Conclusion on compliance and conformance, 30
Confidentiality, 4, 5, 8, 10–12, 95, 107, 109, 114–117, 127, 154, 159–161
Conformance, 28–30, 100, 104, 153
Corrective and Preventive Actions, 99
Create an Inventory of Information Assets to Protect, 115
Cross-Site Scripting (XSS), 107
Cryptanalysis, 9
Cryptography, 1, 9, 10, 131
Cryptology, 9

D

Decipher, 9
Define a Method of Risk Assessment, 114
Description, 52, 57, 60, 65, 69, 75, 77, 83, 86, 110, 117, 120, 131, 134
Determine the Scope of the ISMS, 112
Diagram, 56, 104, 111, 141
Diagrammatically, 52
Disposal, 123, 131, 141, 142, 156, 157, 159–161
Document, 14–17, 19, 20, 22, 33, 35, 36, 61, 65, 67, 86, 87, 89, 90, 92, 93, 97, 105, 108–114, 116, 118–126, 128, 130, 133, 134, 144–148, 153, 158–160
Documentation, 14, 16, 37, 65, 88, 90, 100, 121, 123, 124, 144, 146, 161

E

Elements of information security policy, 13
Emphasizing certain matters without modifying the opinion, 49
Employee, 10, 11, 14–18, 21–23, 28, 43–45, 89, 95, 98, 101, 102, 110, 112, 119, 121, 124, 128, 135, 138, 145, 146, 148–153, 155, 156, 161–164
Enable audit scripts, 74
Encipher, 6, 9
Engineering, 52, 122, 126, 164
Enterprise, 24, 63, 161–163
Enterprise mobility management, 163
Example, 8, 10, 12, 17, 18, 20, 23, 25, 27–29, 44, 45, 50, 52, 54–57, 60, 67, 68, 86–88, 90–93, 98, 100, 103, 113, 114, 116–118, 120, 122, 125, 134–136, 140, 142–144, 146, 148, 151, 154, 160
Example of a well-written nonconformity, 98
Execution, 90, 107
Explain the audit program and the reporting process for deficiencies, 95
Explanation, 39, 46, 47, 55
External confirmation, 90

F

Fabrication, 57
Facility Security Officer, 23

Index 169

Federal, 20, 113, 142, 156
Federal Information Processing Standards (FIPS), 20
Federal Information Security Management Act (FISMA), 20
Federal Information Systems, 20
Fiduciary, 34
File, 37, 60, 61, 67, 68, 71, 72, 75–85, 107, 127, 142, 146, 153, 161
Finance and Audit departments, 20
Finance department, 112–114, 148
Financial, 33, 34, 36, 38, 41, 43, 45–52, 64, 85, 86, 88, 90–93, 126, 135, 140, 144, 146
Fire, 22, 84, 85
Firewall, 148
Firm, 38–40, 64–66, 85, 89
Flow, 5, 52, 56, 57
Frameworks, 15, 19, 20, 50, 63, 64

G

Guidelines, 13, 14, 27, 30, 34, 67, 143–146, 149, 153–155, 158–160, 163

H

How are the CIA and DAD triads mutually exclusive?, 12
How can you relate the CIA triad in your everyday life?, 12
How does corrective action differ from preventive action?, 102
How is corrective action similar to preventive action?, 103
How many domains are there in ISO, 27001?, 131
How many ISO, 27001 Checklists are available?, 59
How to Build an Asset Inventory?, 128
How to find out which ISO, 27001 Checklists are suitable for me?, 59
How to go about it?, 142
How to prepare for an auditor selection process, 63
How to use ISO, 27001 Checklist?, 61

I

Identify Applicable Legislation, 113
Identify Applicable Objectives and Controls, 118
Identify Risks, 115, 116
Importance, 10, 11, 33, 42, 49, 57, 66, 94, 110, 129, 132, 135, 138, 143, 144, 148, 153, 156, 157, 163
Important features of the government audit, 40
Information, 1, 2, 4–10, 13–15, 17–23, 25, 30, 31, 33, 35, 37, 39, 47, 48, 50, 51, 53, 54, 56–65, 75, 78, 84–93, 99, 104, 105, 107–110, 112–119, 121–123, 125–138, 142–161, 164
Information Classification, 123, 131, 153–155
Information Security, 1, 2, 10, 13, 20, 21, 30, 31, 33, 57–62, 107–110, 112–114, 122, 125–133, 142, 145, 147, 150, 154
Information Security and Management System (ISMS), 10, 13, 57–61, 108–115, 118, 122–126, 130
Information Security Overview, 1
Initial password and login settings, 24
Integrity, 5, 7, 10–13, 20, 107, 109, 114–117, 127, 154, 159, 160
Introduction, 10, 13, 30, 31, 33, 91, 94, 134, 150
Introduction to information security policies, 13
ISMS audit checklist, 57
ISMS Purpose and Objectives, 13
ISO, 20, 29, 30, 33, 57–62, 98–100, 104, 107–114, 120, 122–134, 144, 164
ISO 27001, 57, 58, 59, 60, 61, 62, 107, 108, 109, 110, 111, 113, 114, 120, 122, 124, 125, 126, 127, 128, 129, 130, 131, 132, 133, 134, 164
ISO 27001 Checklist, 57–59, 61

K

Key, 10, 13, 25, 28, 36, 47, 60, 63, 66, 70, 72, 76, 84, 95, 97, 127, 131, 138–140, 142, 144, 148, 155, 163

M

Management, 1, 4, 8, 10, 11, 14, 15, 19–22, 28, 30, 33, 34, 36–38, 41, 42, 44, 45, 47, 56–63, 65, 66, 72, 85, 88, 92–96, 99, 100, 102, 104, 105, 107–114, 118, 119, 121–123, 163, 164
Management System, 1, 10, 30, 33, 58–61, 94–96, 104, 107–110, 113, 114, 126, 136, 142
Man-in-the-middle (MITM), 107
Mechanism(s), 1, 2, 4, 6–8, 13, 30, 54, 57
Media, 89, 127, 131, 146, 151, 155, 156, 158–160
Media Handling, 131, 158
Middle, 28
Mobile, 18, 72, 105, 123, 127, 128, 131, 149, 161–164
Mobile device management, 163, 164
Model for network security, 7, 8
Monitor the Implementation of the ISMS, 124

N

Network(s), 1, 4, 6–8, 10, 13, 15, 16, 18, 19, 22–25, 27, 30, 60, 73, 75–77, 81, 105, 112–114, 131, 135, 146–151, 164
Network security, 1, 7, 8, 60, 105, 131
Non-auditable, 59
Non-repudiation, 6

O

Objectives, 13–15, 20, 40, 42, 44, 46, 52, 56, 57, 63, 65, 85, 94, 109, 110, 112–114, 118, 120, 122, 124, 136, 138–140, 143, 144, 154
Observation, 47, 87, 88, 90, 91, 96, 97, 103
Organization, 1, 2, 10, 11, 13–22, 28, 30, 34–36, 38, 41, 43–45, 56–61, 87, 88, 90–95, 98–100, 102–104, 107–110, 112–116, 120, 122, 125–129, 131, 133–140, 142–149, 152–159, 161–163
Organizational, 10, 11, 14, 108, 113, 115, 125, 131, 135, 137, 142–144, 146, 148, 157

P

Passive attack, 2–4
Password, 10, 22, 24, 27, 28, 74, 76, 80–83, 109, 123, 149, 150, 164
Performance, 47, 54, 56, 89–91, 100–102, 135–140, 144
Pervasive security mechanisms, 7
Phased approach, 16
Plan, 14, 33, 34, 36, 37, 52, 53, 56, 61, 63–65, 93, 95, 97, 100–104, 109, 110, 112, 113, 116, 117, 119–125, 130, 135–137, 142, 143, 163, 164
Planned, 37, 47, 111, 124, 130
Planning, 20, 33, 34, 37, 58, 85, 108, 110, 124, 125, 132, 141–143
Policies, 1, 4, 11, 13–15, 17–19, 21, 23, 25, 33, 34, 37, 44–46, 48, 51, 92, 93, 101, 108, 114, 122, 131, 142, 145, 147–149, 156, 157, 163, 164
Policy, 13–20, 22–25, 27, 30, 65, 75, 76, 83, 97, 109, 110, 112, 113, 122–124, 126, 131, 133, 142, 143, 147, 149, 151, 152, 154, 161–164
Policy categories, 19, 20
Prepare/Prepared/Preparing/Preparation, 33–37, 39, 40, 43, 48, 50, 51, 60, 61, 63, 86, 105, 125, 126, 157
Prepare for the Certification Audit, 125
Preventive action, 99, 100, 102–105
Process of Audit Program Management, 36
Profile, 68, 82, 99, 116, 117, 119, 121, 130, 164
Protection, 4, 5, 10, 12, 13, 20, 74, 76, 109, 131, 132, 135, 138, 142–144, 148, 153–158, 160
Purchase a Copy of the ISO/IEC Standards, 110

R

Reasons for Auditing, 33
Remote, 70, 73, 74, 107, 142, 162
Remote code execution, 107
Remotely, 73, 162

Index   171

Responsibility for Assets, 144, 146
Result(s), 9, 10, 12, 19, 27, 29, 38, 43, 46, 52, 54, 55, 78, 84, 88, 92, 93, 101, 102, 104, 109, 110, 112, 113, 116–118, 120, 122–126, 130, 138, 144, 154, 162, 163
Risk assessment, 30, 36, 110, 114, 116–122, 125, 127–130, 133, 142, 160, 161
Risks, 1, 11, 19, 20, 24, 30, 46, 58, 63, 89, 100, 103, 109, 114–118, 120, 122, 126, 127, 129, 130, 133, 136–139, 152, 161, 162

S

Scope, 14, 16, 19, 24, 25, 27, 37–39, 41, 47, 91, 92, 94, 99, 101, 102, 112, 113, 115, 122, 125, 150
Script, 66–85, 151
Scripting, 107
Security, 10–25, 27, 28, 30, 31, 33, 57–62, 67, 68, 70, 75–79, 81–84, 105, 107–110, 112–114, 122, 123, 125–133, 142, 145, 147–151, 153, 154, 156, 157, 161–164
Security attacks, 2
Security Guidelines, 27
Security management policies, 21
Security Mechanisms, 2, 4, 6, 7, 30
Security policies, 4, 11, 13, 14, 18, 19, 131
Security policy audience, 18
Security policy contributors, 16, 17
Security policy development, 14–16
Security procedures, 25, 154, 161
Security Services, 4, 6
Security Standards, 23
Send mail, 25
Server, 18, 21, 23–25, 68, 70–74, 127, 128, 150, 153
Services, 25, 29, 36–38, 42, 44, 57, 58, 61, 65, 72, 73, 75, 84, 85, 105, 127, 131, 146, 149, 150, 159, 163, 164
Set Up Policy, Procedures, and Documented Information to Control Risks, 122
Some basic terminologies, 9
Special applications, 29
Specific security mechanisms, 6
SQL Injection, 107
Standard, 10, 14, 16, 20, 21, 23–25, 27–30, 33, 34, 38, 46, 47, 50, 55, 66–68, 88, 92, 95, 97–100, 104, 108–110, 113, 118, 120, 124, 130, 148

T

The CIA and DAD Triads, 11, 12, 107
The CIA triad, 11, 12, 107, 108
The DAD triad, 12
The OSI Security Architecture, 1
Techniques, 7, 8, 10, 44, 57, 63, 86, 88, 125, 159
Triad, 11, 12, 107, 108

U

Unified Endpoint Management, 163

W

Web, 18, 25, 60, 70, 84, 126
Web server, 25, 70
What is the basis of the ISO, 27001 Checklist, 61
Who all can use ISO, 27001 Audit Checklist?, 58
Who has prepared ISO, 27001 Checklists?, 61
Who has validated ISO, 27001 Checklists?, 61
Who Should be the Asset Owner?, 128
Why security management?, 10

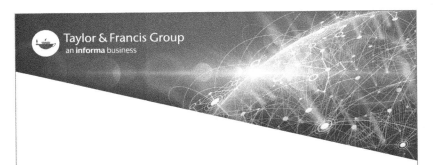